Designing Knowledge Economies for Disaster Resilience

Catastrophes in Context

General Editors:
Robert E. Barrios, University of New Orleans
Crystal Felima, University of Kentucky
Mark Schuller, Northern Illinois University

Catastrophes in Context aims to bring critical attention to the social, political, economic, and cultural structures that create disasters out of natural hazards or political events and that shape the responses. Combining long-term ethnographic fieldwork typical of anthropology and increasingly adopted in similar social science disciplines such as geography and sociology with a comparative frame that enlightens global structures and policy frameworks, *Catastrophes in Context* includes monographs and edited volumes that bring critical scrutiny to the multiple dimensions of specific disasters and important policy/practice questions for the field of disaster research and management. Theoretically innovative, our goal is to publish readable, lucid texts to be accessible to a wide range of audiences across academic disciplines and specifically practitioners and policymakers.

Volume 7
Designing Knowledge Economies for Disaster Resilience: Case Studies from the African Diaspora
Edited by Pamela Waldron-Moore

Volume 6
The Power of the Story: Writing Disasters in Haiti and the Circum-Caribbean
Edited by Vincent Joos, Martin Munro, and John Ribó

Volume 5
Making Things Happen: Community Participation and Disaster Reconstruction in Pakistan
Jane Murphy Thomas

Volume 4
Constructing Risk: Disaster, Development, and the Built Environment
Stephen O. Bender

Volume 3
Going Forward by Looking Back: Archaeological Perspectives on Socio-ecological Crisis, Response, and Collapse
Edited by Felix Riede and Payson Sheets

Volume 2
Disaster Upon Disaster: Exploring the Gap Between Knowledge, Policy and Practice
Edited by Susanna M. Hoffman and Roberto E. Barrios

Volume 1
Contextualizing Disaster
Edited by Gregory V. Button and Mark Schuller

Designing Knowledge Economies for Disaster Resilience

Case Studies from the African Diaspora

Edited by Pamela Waldron-Moore

berghahn
NEW YORK • OXFORD
www.berghahnbooks.com

Published in 2024 by
Berghahn Books
www.berghahnbooks.com

First paperback edition published in 2026

Library of Congress Cataloging-in-Publication Data

Names: Waldron-Moore, Pamela, editor, author.
Title: Designing knowledge economies for disaster resilience : case studies from the African diaspora / edited by Pamela Waldron-Moore.
Description: New York : Berghahn Books, 2024. | Series: Catastrophes in context ; Volume 7 | Includes bibliographical references and index.
Identifiers: LCCN 2023021758 (print) | LCCN 2023021759 (ebook) | ISBN 9781805391708 (hardback) | ISBN 9781805391715 (ebook)
Subjects: LCSH: Knowledge economy—Developing countries. | Economic development—Environmental aspects—Developing countries. | Natural disasters—Economic aspects—Developing countries. | Natural disasters—Planning—Developing countries. | African diaspora.
Classification: LCC HD30.2 .D4677 2024 (print) | LCC HD30.2 (ebook) | DDC 658.4/038—dc23/eng/20230905
LC record available at https://lccn.loc.gov/2023021758
LC ebook record available at https://lccn.loc.gov/2023021759

British Library Cataloguing in Publication Data
A catalogue record for this book is available from the British Library

EU GPSR Authorized Representative

LOGOS EUROPE, 9 rue Nicolas Poussin, 17000, LA ROCHELLE, France
Email: Contact@logoseurope.eu

ISBN 978-1-80539-170-8 hardback
ISBN 978-1-80758-142-8 paperback
ISBN 978-1-80539-400-6 epub
ISBN 978-1-80539-171-5 web pdf

https://doi.org/10.3167/9781805391708

An electronic version of this book is freely available thanks to the support of libraries working with Knowledge Unlatched. KU is a collaborative initiative designed to make high-quality books Open Access for the public good. More information about the initiative and links to the Open Access version can be found at knowledgeunlatched.org.

For my daughter, Farah, my grandson, Micah, and all students of the African diaspora, without whose desire for knowledge of the past and hope for the future, ideas for this book would not have germinated.

Contents

List of Illustrations viii
Acknowledgments x

Introduction 1
Pamela Waldron-Moore

Chapter 1. Knowledge Production for Disaster Recovery 7
Pamela Waldron-Moore

Chapter 2. Conceptual Framework: Knowledge Capital in the African Diaspora 25
Pamela Waldron-Moore

Chapter 3. Louisiana: Collaborative Energies and Post-disaster Solutions 65
Jessica Dandridge and Yeishka Montalvo

Chapter 4. Haiti: Building Social Capital through Media Connectivity 105
Shearon Roberts

Chapter 5. Rwanda: Sociocultural Practices for Resilience and Recovery 130
Samuel Habimana and Emmanuel Biracyaza

Chapter 6. Sankofa: A Model of Sustainable Development 159
Pamela Waldron-Moore

Chapter 7. Epilogue: Post-disaster Solutions for Development 194
Pamela Waldron-Moore

Appendix I 212
Appendix II 214
Index 219

Illustrations

Figures

1.1. Knowledge Economy Word Cloud. © Pamela Waldron-Moore. 11

2.1. Essential Pathways to Knowledge Economies, Conceptual Framework. © Pamela Waldron-Moore 2023. 28

2.2. Black Households and Broadband Access in Louisiana (2018 American Community Survey included in the Louisiana Budget Project) Data taken from: https://www.labudget.org/2020/05/separate-and-unequal-students-access-to-technology-in-the-time-of-coronavirus/. 47

2.3. Real GDP growth in Ghana 1980-2028. Data from: https://www.imf.org/external/datamapper/NGDP_RPCH@WEO/GHA?zoom=GHA&highlight=GHA. 53

3.1. Mapping Engaged Louisiana Community. © Pamela Waldron-Moore 2022. 98

4.1. #PetwoKaribe Movement. Mobilizing Social Capital Both Online and Offline (Mirambeau 2018) | #PetroKaribe Movement. Educating and Raising Awareness on Post-Disaster Sustainable Recovery (IJDH 2018). 107

4.2. Ayiti Kale Je project. Word Cloud generated by Shearon Roberts using TagCloud application. 117

5.1. Socio-therapeutic training exercise. Source: RRGO project © Samuel Habimana. 136

5.2. Gathering of genocide survivors, perpetrators, and family members. Unity and Reconciliation program. © Samuel Habimana. 139

5.3. Government provided green space available for Umuganda activities. © Samuel Habimana. 144

5.4. Ndi Umunyarwanda Gathering. © Samuel Habimana. 148

6.1. The Sankofa. Mythical African Bird. Wikimedia Commons, public domain. 161

6.2. Global Wealth Distribution (Property) 2020 Data Source: Credit Suisse Research Institute, Global wealth report 2021, June 2021. Chart created by Leandro Salvador for public domain: https://en.wikipedia.org/wiki/File:Global_Wealth_Distribution_2020_(Property).svg. 167

6.3. Global North Gender Gap, 1970-2016. Source: https://ourworldindata.org/grapher/gender-wage-gap-oecd?country=AUS~FRA~ITA~JPN~SWE~GBR~USA. 171

7.1. Louisiana. Pillars of Collaborative Intellectual Energy. The Water Collaborative (TWC). Developed by Jessica Dandridge. 201

Tables

6.1. Race-Gender Wage Comparisons in Louisiana. © Pamela Waldron-Moore. 165

6.2. Costs, Benefits, and Outcomes of Global Climate © Pamela Waldron-Moore. 183

Acknowledgments

PAMELA WALDRON-MOORE

Climate change is fast becoming an existential crisis for the universe. It is, however, more negatively impactful for the African diaspora, whose vulnerabilities are specifically related to low economic development and global injustice. Acknowledging the political, economic, and social foundations of the problem, it is important to identify the challenges and opportunities facing the diaspora. Teaching at a historically black university in the southern United States and experiencing the almost annual trauma of weather disasters, I am reminded of an old African proverb that counsels us: "if you want to go fast, go alone, but if you want to go far, go together." The urgency to mitigate climate disaster requires educators collectively to motivate and empower the next generation to develop their intellectual space in the universe of ideas and scholarship to mitigate climate risks, recover from the traumas of disaster, and create sustainable futures for their communities. The paucity of educational resources and the relative absence of knowledge sharing across communities to assist in this process informs the writing of this book.

Although this venture only took shape in the last two years, I have written and discarded several prior attempts in my head, in brief unpublished pages of my to-do portfolios, in presentations submitted at pedagogical forums, and podcasts hosted by my institution's center for the advancement of teaching. To produce a clear and coherent study in the discipline of political science, where democratization of thought and the globalization of knowledge have left little room for shining a torch on those left behind in the stampede to progress and economic advancement, seemed an insurmountable task for the time and effort this book required. Yet, it was a task I was determined to address before the candles of those passions flickered out.

The completion of this manuscript has left me indebted to organizations and individuals who have made space in their tightly packed calendars to accommodate access to library resources, the reading of early drafts, and referrals to individuals engaged in conversations relating to the content of my work. Indeed, were it not for the intrusion of the world's most recent disaster, the COVID-19 pandemic, which gave many of us windows of time we would not have anticipated three years ago, this effort would not have been forthcoming. Therefore, I want to recognize personnel at New York University's Elmer Holmes Bobst library who allowed me virtual and in-person access to library holdings on disaster management while a scholar-in-residence there, an effort funded by UNCF/Mellon under the sponsorship of Dr. Cynthia Neal Spence, Director of UNCF/Mellon Programs and Sociology Professor at Spelman College.

I also want to acknowledge the many online libraries helpful in researching literature on climate disaster and the diaspora. Through these resources, I was reminded of the work of the green belt movement, inspired by the late Wangari Mathai, Nobel Peace Prize winner, whose consciousness on what Africa and the diaspora could do through community empowerment for sustainable development reinforced the thrust of this manuscript. Gratitude is also extended to Aubrey Paris, host of the innovation station series on gender, climate, and innovation, sponsored by the State Department's global women's issues initiative, which introduced me virtually to the innovative spirit of women in the Gulf of Mexico and other regions through transformative discussion on disaster mitigation and related action to achieve environmental sustainability. Virtual access to additional online libraries, provided through Xavier University of Louisiana, has also been instrumental in facilitating research on disaster resilience and recovery.

I especially want to thank, however, three social science editors of the Catastrophes in Context series, two anonymous peer reviewers, and the editors at Berghahn Books, without whose critiques and guidance, respectively, this book could not have been successfully concluded. The patience of Professor Jeff Jensen, a political scientist serving at New York University's satellite institution in Abu Dhabi, who made the time to read early drafts of the book and generously commented on its viability and quality, is much appreciated. A wealth of gratitude is also owed to two of my former students, Jessica Dandridge and Yeishka Montalvo, whose activism on climate justice contributed important insight on Louisiana; my colleague, Shearon Roberts, whose scholarship on Haiti has added meaningful context on knowledge production to this book; and two Rwandan doctoral students, Samuel Habimana and Emmanuel Biracyaza, whom I had the pleasure to learn from, beginning with a trip to Rwanda in 2019 on the occasion of the twenty-fifth commemoration of the Rwandan genocide, and culminating in a witness to the resilience of the African psyche and shared in the chapter on Rwanda included in this volume. Along with other friends and scholars exploring themes of disaster, injustice, underdevelopment, and sustainability, I hope this work reflects the joint aspiration of students of the diaspora to awaken to new innovative approaches to bottom-up community efforts in disaster recovery.

Finally, I offer my heartfelt thanks to all the students at Xavier University of Louisiana who, over the course of twenty-five years, have inspired me with their intellectual curiosity and eagerness to learn about knowledge capacities that may have eluded them. Their collective energies have propelled me to the conclusion of this work. Especially to a most ardent mentee, Jailyn Williams, a student of political science and African American and diaspora studies, who has meticulously proofread and commented on every aspect of this book from a student's perspective, I am truly grateful. And last but not least to my daughter, Farah, and others who have not been named in these acknowledgments but have tangibly or intangibly, in ways both small and large, contributed to the production of this effort, I remain deeply indebted.

Introduction

PAMELA WALDRON-MOORE

Over the past thirty years, disaster research has exploded into focus, highlighting its relationship to climate, political, and social change and its global impact on economic development and sustainability. This research has brought into sharp contrast, the question of climate justice and the widening gap between those who are able to engage in risk mitigation and adaptation to greener infrastructure via just transitions for the sake of a healthier world and others who want to do so but simply cannot. In his *Theory of Justice*, John Rawls (1971) sought to extend the idea of fairness to international law for the purpose of judging the aims and limits of a just war as it applies to international communities of human beings. He referred to the latter as well-ordered peoples, many of whom are left out of consideration, because they are denied a meaningful role in political decision-making (Rawls 1999: 23–43). According to Rawls, the most important way to think about justice is to determine what principles would be chosen by people who came together behind a "veil of ignorance" that temporarily deprived them of any knowledge of where they would find themselves in society. The point is that justice is only likely to be achieved when there are no differences in bargaining power or knowledge. This seems to represent well the imbalance currently observed in actions of the international community on issues such as climate justice.

In 2019, Greta Thunberg, a teenage Swedish environmental activist, in her address to the United Nations Climate Action Summit, drilled home the lack of fairness across the international community by deriding world leaders for focusing on the potential costs of climate change in the face of industrial advancement and challenging leaders to take immediate action to mitigate its risks rather than seeking to increase their economic growth without identifying and sharing ways to balance global development equitably. The single most critical point of her argument was the lack of knowledge of so many leaders and publics of the efforts necessary to preserve our world in the face of accelerating climate crisis or the lack of will among those who know what is needed but fail to take meaningful action. Without equitable access to knowledge or a seat at decision-making tables, vulnerable communities may not ever experience an equitable balance in global development.

This volume is intended to contribute to the conversation on the growing gap in awareness of the world's peoples to the concept of disaster by investigating the nature of disaster, the impact of disaster, and the role of knowledge production and knowledge sharing in mitigating losses that occur as a result of inaction on the part of a global community either too crippled to act fairly or lacking the knowledge to do so. Especially in underdeveloped spaces, which include Africa and its diaspora, there is need to devise a model of regenerative development that grows from the bottom up rather than applying the top-down extractive methods they have been cajoled into accepting as the only developmental pathway forward. Because the Global South has experienced more than its fair share of disaster and less than its fair share of the world's wealth, we will highlight areas of disaster and the resilience with which impacted communities have survived and have the capacity to prevail sustainably.

Organization of the Book

Contributors to this book have engaged in extensive research on various aspects of climate and other disasters, with an eye to chronicling post-disaster experiences that may be helpful to less developed countries that have not been included in decision-making forums at the global level; nor have they been able to garner adequate support from global agencies engaged in research on disaster mitigation. With about twenty-five years of combined research as educators and students, the authors have identified the location of knowledge capital, the application of knowledge production, and the innovative patterns of knowledge sharing as a model designed to achieve disaster resilience and economic development as a viable pathway for managing post-disaster trauma in disadvantaged spaces of the Global South.

Chapter 1 explores the underlying propositions of this book. It focuses on the knowledge base required to identify the capacities of Africans in the diaspora, the location of indigenous knowledge, and ways in which knowledge capital may be produced to create a just transition to disaster recovery, economic development, and environmental sustainability. It stresses the belief that without an adequate model for development, emerging nations and especially those of the African diaspora, a community of people exploited for centuries and deliberately omitted from the lexicons of most databases and archives recounting the disastrous episodes of their lives, there can be no achievable growth. Thus, identifying new ways of knowing, building on the indigenous knowledge of their past, and converting native knowledge capital into knowledge economies, a sustainable future may be designed—a future that transports the past into the present and prepares to forge a resilient path forward.

Chapter 2 offers a conceptual framework of how knowledge capital may yield development and sustainability once the capacity of communities is identified and produced. It is aimed at presenting a theoretical overview of the problems for development in pre- and post-disaster communities and explores the philosophy that when communities draw on their indigenous cultural experiences, they are more likely to reclaim perspectives of self-determination and development rather than continue to flounder under pressure to imitate the West in a linear, exogenous path to development. When lock-step political behaviors fail to yield economic advancement in a globalized world and dependency on the nurture of colonial masters does not work, national self-interest requires disadvantaged regimes to find strength and direction in proven cultural traditions through the exploration of strategic intellectual capital.

Engaging systems theory (Easton 1965), this chapter demonstrates the role of knowledge capital at its source in disadvantaged communities where disaster has forced communities to address alternative, indigenous sources of resilience to rebuild a nation. In Easton's work, he defines the political system as a set of interactions, abstracted from the social behavior, through which values are authoritatively allocated for society. This rendering of the political system places the government at the center, receiving inputs/demands from the public and responding to those demands with decisions and actions that will regulate outputs for the society. Thus, this earlier model needs to be retired in order to empower community building, reliant not on individual government actions but on an ecosystem of networks in which justice and solidarity create a better framework for growth.

Chapter 3 is a case study that examines collaborative energies and post-disaster solutions for development in Louisiana. The chapter highlights the inherent value in developing solidarity networks to fight for a world in which dignity, self-determination, and a sustainable future becomes a reality. Emerging post-disaster societies in the Gulf of Mexico and the Caribbean Sea have found community capital a meaningful way to progress in pursuit of new trajectories for growth and sustainability. The Gulf South is a region where water, oil, gas, coal, and nuclear energy are key segments of the economy. Louisiana communities, particularly the black, brown, and indigenous people, are becoming increasingly vulnerable to the subsequent land loss, rising seas, natural disasters, crumbling infrastructure, and polluting energies weakening an already fragile ecosystem. The climate crisis has compounded vulnerabilities and become a turning point for solidarity, urging local communities to act, connect their struggles, and overcome the systemic inequalities that divide them. Movements and coalitions are united to create a resilient and sustainable future for all, and their fight is anchored in the knowledge capital, histories, realities, and power of the Gulf region. By advancing grassroots policy and practices that center laborers, farmers, fisher-folk, tribal nations, and frontline communities in a just transition away from extractive economies, sustainable development can be attained.

Chapter 4 reflects on Haiti and its capacity for building social capital for sustainable growth via media connectivity. Although Haiti's recovery efforts have been stymied by continuing issues of governance and the costs of development, there is a grit among Haitians that can be awakened to a new search for development. This chapter pinpoints the fact that the public sphere is central to the collective shaping of local ways of knowing. Through citizen engagement in dialogue, debates, advocacy, and even creative forms of expression, societies propose and refine notions of ways of life and uphold commonly shared values of humanity and dignity. Media platforms, both traditional and digital, offer forms of public spheres that transmit public knowledge, stimulate debates on the status quo, and serve as a rallying call for action in the face of persistent political and social stalemates.

In Haiti, traditional media, both commercial and community radio, played an important role during the 2010 earthquake and the decade since. In its quest for a more equitable recovery and sustainable way of life for ordinary Haitians, digital platforms have bolstered local and global conversations that aim to hold those in power accountable for unsustainable forms of governance that continue to increase hardships for Haitians. Similarly, the creative genius of Haitians, overseas and at home, have relayed stories of the past and the will of ordinary citizens to devise ways of overcoming insecurities in pursuit of survival and a better life. To help Haiti take its best shot in designing a happier future than it has experienced in the face of recent political, social, natural, and man-made disaster, the awareness must be shared that the success of their lives lies in the resilience of their spirits and the knowledge capacity they own. Capitalizing on these will craft a mosaic of achievement for the future.

Chapter 5 addresses socio-cultural practices for resilience and recovery in post-disaster Rwanda. The genocide in Rwanda (1994) was a catastrophe of enormous proportion. It left in its wake a huge loss of human life, unparalleled socio-cultural distrust among communities, and a chasm in political legitimacy that has only, during the last decade, begun to show promising signs of closure and a clear track toward sustainable development. The Rwandan people have been hailed by the media and economic reviewers for their resilience throughout the twenty-eight-year period of rebuilding. Although the strides made have been supported by collaboration with European and other counterparts, the methodologies employed to aid recovery in Rwanda were generated by local communities working to sustain development from the community level upward. To heal the nation and return to a semblance of harmony among groups formerly hostile to each other, integrative psychotherapeutic approaches have been adopted to reflect traditional Rwandan culture, assert respect for personal loss within communities, and dispose of victims in a traditional, dignified, and respectful manner. In these and other ways, Rwanda has been a lighthouse for managing trauma and restoring justice in their post-disaster period. The post-disaster experience of Rwanda may empower states in the African diaspora to seek new and innovative ways to overcome decades of trauma they have collectively endured.

Chapter 6 presents a model of Sankofa beliefs for tackling sustainable development goals in the African diaspora. Sankofa is a bird in African mythology with its head turned backward and its feet planted firmly forward. Descendants of the Akan people of Ghana, many trans-shipped to the Americas and Caribbean, understand this as a metaphor for employing knowledge of the past as a motivation for bringing wisdoms of the past into the future, for growth and development. In the context of climate change and environmental disaster, the greatest global challenge of the twenty-first century, one recognizes the symbolism of Sankofa as conceptualizing knowledge production. The diaspora's wisdoms lie in its intellectual past, which may be honed to produce an innovative, empowered present, capable of creating a viable and sustainable future.

Finally, the Epilogue of this volume, reiterates the view that adaptation is at the heart of disaster mitigation in the three diasporic sites highlighted in these case studies. Insights that contribute to a favorable outcome for the societies in this study may be applied to pre-disaster conditions as well as to post-disaster strategies to seek a trajectory of development. This emanates from the identification of intellectual capital resident in these societies and capable of producing innovative methodologies for potential sustainability. As knowledge capital is produced, the achievement of goals will empower communities to generate conversations that will contribute to a feedback loop that energizes even more innovative ideas for tackling economic insecurity, gender inequality, and environmental injustice. This volume will ultimately evaluate the impact of a secure economy, gender equity, and climate justice on sustainable development. Reflecting on the past may remind Haiti and Louisiana of commonalities in their relationship and probable reasons for reconnecting to shared history and potentially investing in shared strategies for building a resilient future.

Conclusion

It is expected that academic institutions, especially historically black colleges and universities (HBCU), will not only find this volume instructive for students and faculty but will also use it to arouse student interest in innovative projects that may be shared with the diaspora in pursuit of sustainable and developed communities to empower and advocate for them in the management of sustainable post-disaster societies. Publication of innovative work conducted by faculty and students in response to the discussion in this text will further advance our attempt to awaken a spirit of awareness and solidarity among diasporic peoples and a will to build an edifice of knowledge in which access to collaborative knowledge sharing may be forthcoming. In the spirit of Wangari Maathai (2003), a Kenyan social, environmental, and political activist and the first African woman to win the Nobel Peace Prize (2004), the diaspora must mobilize as a unified indigenous diasporic grassroots organization that empowers its communities to address imperatives for post-disaster sustainability. Only in such collaboration is there a realistic chance of environmental survival in Africa and its scattered diaspora.

References

Easton, David. 1965. *A Systems Analysis of Political Life*. New York: John Wiley.
Maathai, Wangari. 2003. *The Green Belt Movement: Sharing the Approach and the Experience*. New York: Lantern Books.
Rawls, John. 1971. *A Theory of Justice*. Cambridge, MA: Harvard University Press.
Rawls, John. 1999. *The Law of Peoples*. Cambridge, MA: Harvard University Press.
Thunberg, Greta. 2019. "Presentation at the United Nations Climate Action Summit," 23 September 2019. New York: United Nations.

CHAPTER 1

Knowledge Production for Disaster Recovery

PAMELA WALDRON-MOORE

The African diaspora constitutes the world's collection of communities with people descended from Africans and residing all over the globe but, perhaps, predominantly in the Americas and Caribbean, following the mass dispersion from Africa during the transatlantic slave trade of the 1500s–1800s. Most of the African diaspora, also referred to as the Black diaspora, were dispersed from West and Central Africa but maintained their cultural traditions throughout the period of and well after their enslavement. Interaction with the indigenous and other communities and the rest of the world has contributed to cultural legacies influenced by Europeanization, geography, psychological battery, learned resilience, a strong sense of community, etc., and passed on through the generations. In the post-emancipation period, some of the lessons learned in the past about survival and resilience have also given way to migration, dependency, and experiences of inequity, marginalization, and poverty.

Knowledge production in Africa, and passed down through its diaspora, stems from unique ways of knowing that are grounded in indigenous African cultural knowledge systems (Mpofu, Ntinda, and Oakland 2012). Yet, despite the diversities in language and culture, there are enduring commonalities that remain rooted in indigenous African traditions, ecology, and history (Ngara 2007). Colonial rule, especially in the former British colonies, where assimilation was not forced upon Africans as was the case under French colonial systems, allowed Africans to embrace both their indigenous cultural heritage and adopt Anglo-Western knowledge systems (Ngara 2007). Thus, regardless of where Africans were dispersed, it seemed evident that the two knowledge systems enriched *knowing* across the diaspora and may have also contributed to the perception of scholars that Africans across the diaspora experience unity in diversity (Goduka

1999). This volume hopes to build on the commonalities and sense of unity in diversity understood by African-descended people as a resilience that defines who they are as a people and how their worldview is shaped.

Indigenous ways of knowing, described by some scholars as the African paradigm (Ngara 2007), have been ridiculed, misunderstood, misinterpreted, and rejected by colonizers over the centuries. Both the continent of Africa and its people have been treated as uncivilized and without a cultural mind. These flawed observations were seen as justification for subjugation and encouragement to tame the savage within them, to denigrate their knowledge systems, and strip them of all indigenous systems of survival. Cultural hegemony, once launched on Africans, served to rid them of their self-esteem and political efficacy and throw them into a dependence syndrome that is today, mostly responsible for the fact that Africans at home and in the diaspora still rely too much on the colonizing forces of the past and the economic structures they set in place with African labor and culture at their base (Goduka 1999).

In recent decades, climate change, with its impact on development, has brought the concerns of Africans in the diaspora to the fore, with many seeking to change the narrative of a proud but disadvantaged people, to reflect the challenges and opportunities facing the diaspora and the resilience with which so many have overcome the obstacles before them. Especially in disaster-prone spaces, the necessity to overcome the obstacles of climate change and the economic deprivation experienced over centuries must rely on alternative approaches to survival and sustainability. Thus, we are recommending, for the diaspora, a shift from sole reliance on European benevolence and foreign-born technology to knowledge sharing of intellectual capital, cultural wisdom, and homegrown solutions, focused on nature-based rather than on man-made solutions. Emphasis is placed on seeking a just transition to food and economic security, limitations on coastal erosion and soil degradation, and a purposeful pursuit of environmental justice in housing.

We call for neighborhood locations to be replaced by green infrastructure and sustainable development via easy access to education and technological expertise. Such education must emanate from a return to indigenous ways of knowing with which new material is integrated to advance communication in a blend of cultural knowledge and value systems. As Sol Obotetukudo (2001) argues, development in Africa and its diaspora cannot be realized without an African philosophy of development derived from what Africans think of themselves as informed by their indigenous cultural knowledge. Such cultural knowledge is typically passed down through generations in African proverbs, songs, and music. Indeed, true development will only take place and become sustainable when African culture and value systems are shared among them, in diasporic spaces, and globally. This volume seeks to add to the paucity of literature on this subject to shed light on how multilayered systems of knowing and the production of new knowledge may contribute to meaningful learning (see Ausubel 1963) that may inspire post-disaster communities to realize progress and development in a model of knowledge production.

Knowledge Production for Disaster Recovery

Disaster research has been studied from a variety of angles, often taking a historical approach or contemporaneous reporting as new crises occur. Seldom has there been targeted examination of the implications of disaster for vulnerable societies. Most developing societies are vulnerable to the vicissitudes of climate change but none more so than communities across the Black diaspora. Unlike emerging societies in Asia, which may also face the challenges of climate change and negative impacts to their development, Africa and diasporic communities have not developed a model for economic growth that does not depend on fossil fuels and energy-based industrialization for environmental sustainability. As we may recall, the East Asian model for development is built on manufacture and export-oriented industry, crafted under the cultural belief that command economics is a viable approach (Park 2002). For the relatively short period of political independence, the diaspora has maintained strong ties to the colonial powers who continued to extract from them their abundant natural resources. In the wake of disaster, created by natural hazards or cultural or political events, which continue to exploit natural resources to the disadvantage of development, there is need for leaders of diasporic states to scrutinize the social, political, economic, and cultural structures that have inhibited development and stagnated the advancement of self-determination in small, developing, post-disaster societies.

This volume interrogates the challenges for disaster-prone territories of the African diaspora to determine supplemental strategies that may be embraced to effect policy changes that might contribute to sustainable communities. Acknowledging the link between low economic development and climate change to be one of codependence and nonviability, we must come to terms with the fact that despite public and private capital flows into diasporic and other developing communities, the latter are unable to locate enough financial resources to support green energy projects, as recommended by the United Nations (Edwards 2019; UN 2020) and its advisory body, the International Panel for Climate Change (IPCC). The damage to diasporic communities experiencing negative impacts from climate change constitutes a catastrophe of enormous proportions (Gallagher 2022).

The prospect of mitigating the risks of enhanced disaster to these vulnerable communities warrants a model of economic growth that balances innovative energy with community resilience. As global warming persists and northern climates become wetter while southern climates become dryer, there is no time to lose and certainly no time to continue dependence on the advanced, industrialized societies, either to right the wrong of impoverishing developing societies with extractive techniques, which have resulted in damage to the atmosphere, or in providing the kind of financial and technological assistance they would need to jumpstart flagging economies. In proposing the creation of knowledge economies by identifying ignored pockets of intellectual capacities resting in homegrown expertise, communities of the diaspora must recognize innate abilities to help themselves via identification of knowledge capital, innovative production of said capital, and advocating for growth in communities while empowering local leaders to engage in regenerative enterprises for sustainability.

The knowledge economy is an economic system in which the production of goods and services is based primarily on knowledge-intensive activities that contribute to growth in technological and scientific innovation. Knowledge economies, as captured in the word cloud below, recognize growth to be dependent upon the quality and quantity of information available to communities and permitting access to sources of information. Locating available information is more critical than the means of producing information at the grassroots level. Yet the production of knowledge acknowledges the importance of locating tangible and intangible values stemming from the identification of local data about processes, products, customs, and other intellectual capacities that allow for efficient ways to coordinate intellectual capital and creative capacity. Not only must knowledge production be recorded for easy access when necessary, therefore, it must also be shared for replication and in self-help activities. As claimed by Justin Rosenstein (2012): "In a knowledge economy, natural selection favors organizations that can most effectively harness and coordinate collective intellectual energy and creative capacity."

Knowledge sharing in the past has been a beacon for the present. In the age of Neanderthals, for example, lessons were shared via cave drawings. Tool-making, animal hunting, and other endeavors were documented in hunting expeditions, illustrating best practices for survival skills. Through such methods, a knowledge base of day-to-day activities was created. However, the intellectual capital and capacities of Africans in the diaspora have over the ages been co-opted and silenced, rendering access to via-

Figure 1.1. Knowledge Economy Word Cloud. © Pamela Waldron-Moore.

ble knowledge bases relatively difficult to locate. As demonstrated in the appendices to this volume, communities working together can build on the few knowledge bases recording the innovative spirit of Africans in the diaspora by managing the knowledge capacities resident in their locales to explore what actions, taken in the past, may be re-designed to address post-disaster needs of the present for the future. In today's environment, analysis shows that interaction among working groups identifies knowledge sharing as 28 percent in the area of reading and answering email messages; 19 percent in searching and gathering information; 14 percent in internal communication and collaboration; and 39 percent in role specific tasks (International Data Corp 2022). This representation is relative to developed areas of the Global North but reflects very little of what takes place in post-disaster societies where there is a gap in information dissemination to regions of the Global South.

Diaspora Considered

Three diasporic sites that enhance investigation of our theme, and are explored in this volume, are Louisiana, Haiti, and Rwanda, where lessons learned from disaster may be shared and their resident opportunities uncovered. Louisiana, located in the developed world but with the second largest (33 percent) population being of African descent, and Haiti, with 95 percent of its population being of Black origin, have both experienced the vulnerabilities of physical disaster, coupled with limiting inequities, that have hindered, in critical ways, progress toward sustainable development. Rwanda, located in Central Africa, has experienced political and cultural disaster and is moving slowly, but resolutely, toward recovery by harnessing the knowledge capital in communities whose native courage and resiliency have been recounted in anthropological studies and traditional practices. We believe that knowledge sharing within and across the diaspora could serve as a template for recovery and empower sister-communities to pursue new approaches to economic development. Innovative engagement of homegrown practices, long neglected in the face of contemporary global opportunities, may hold promise for achieving environmental sustainability in the foreseeable future.

Analysis in this study engages directly with the wisdoms of indigenous practices in Africa and the diaspora to craft new pathways for development in the wake of national disasters that have stagnated growth in their societies. These indigenous practices will be discussed in detail relative to the themes under consideration. The areas of development primarily targeted in this volume are those that have been recognized globally as risk factors. For example, water management, poverty, and public health concerns are high on the UN's goals for sustainable development. Although, as Kelly Gallagher (2022) reported in a recent *Foreign Affairs Newsletter*, the International Energy Agency (IEA) has estimated it would take over $4 trillion in annual investments in clean energy to decarbonize the global energy system, negotiators have come nowhere close to realizing such a bold sum. And, even if they were to raise such funds, what would be shared with the developing world would be next to nothing, considering that those who could advance the funds for global expenditure will spend the lion's share on their own domestic climate needs.

So, what is to become of diasporic locations whose populations need to find ways to recover from disasters caused by climate change and are much more urgent than the climate challenges facing industrialized societies? Through the lens of equity, development indicators, especially those measured by gender inequality, climate injustice, and economic insecurity, which are all included in the seventeen bullets highlighted in the UN Goals for Sustainable Development (UN 2020), we will observe, qualitatively, the potential impact of knowledge production for achieving developmental sustainability in disadvantaged and disaster-prone spaces of the African diaspora, with a view to sharing the successes achieved and innovative strategies applied for achieving them, with other diasporic locales.

Historical Reflection

Omitted from general discourse on development has been the role individuals can and do play in policy-making for development. The integration of national economies through the process of globalization has advanced the notion that growth can only be achieved and sustained through the application of high tech and manufacturing expertise. Global North countries have exemplified the adequacy of this theory by extracting natural resources from Global South states to convert them, with technology, into manufactured goods. This has created a lag in wealth and development between the North and South. Through this prism of global development, advancement is singularly economic and almost entirely tech-based. The reality is that national development is not unidimensional. Emerging nations, and especially those experiencing post-disaster trauma, need to foreground other integral aspects of growth as seen at the intersections of their social, cultural, political, and economic histories. Such reflection may be more likely to support a creation of knowledge economies as a model for growth.

Our volume argues that when the above factors are taken into consideration, there is stronger motivation to push beyond the well-known prescriptions for development to achieve the kind of growth affected communities want to experience. Post-disaster societies may benefit from developmental options relative to their own cultural and scalable spheres of reference. Supplementing the tech-based prescriptions of advanced economies with novel areas of production can jumpstart a return to cultural wisdom with the development of innovative ideas rooted in untapped indigenous knowledge capital. The Black diaspora has lost so much by way of knowledge. Much of the accomplishments of Black people have been buried with those who knew what was accomplished but were often too modest or too low in self-esteem to share their *knowing* or lacking in access to vehicles for knowledge sharing. In 1905, W. E. B. Dubois made what is today remembered as the Niagara Movement Speech. In his presentation, he argued that "either the United States will destroy ignorance or ignorance will destroy the United States" (DuBois 1905). It is hard not to see why that sentiment still resonates in our educational systems, where students of the diaspora continue to be discouraged from learning about their ancestors and the resilience among them that has paved the way for new empowering scholarship in contemporary movements such as Black Lives Matter and in the writings of Black anthropologists who are still too few in number to investigate culture across the diaspora. For, although Black inventors are uncommon in developed locations, some of their work can be identified and correctly attributed to them (Dass 2020).

That is not paralleled in Africa, the Caribbean, or Latin America, where many are credited for their cultural contributions in art, sculpting, and so on but few are acclaimed for technological innovations on par with the creation of traffic signals or refrigerated vehicles or automatic elevator doors, all of which have been contributions made by African Americans (see Appendix I and II). Connecting the Black diaspora through knowledge sharing can be empowering when knowledge economy is embraced as a viable model for economic growth and development. The irony of this is that people of Africa and its diaspora are content to believe in their former and current colonizers. As one of their most vocal sons, Professor P. L. O. Lumumba (2022) has claimed in one of his many educational speeches, Africa is politically and economically weak, socially disorganized, and culturally confused. What was even more disturbing, but encouraging at the same time, was the assertion that in Africa, inter-African trade makes up only about 15 percent of trade, there is no export in agricultural products, no production of African technology in areas, such as mobile phone production, even though the technology for said production is located in the Congo, and no development of pharmaceutical products, a finding recently exposed during the COVID crisis when all of Africa was relying on China, Britain, and the United States for vaccine production.[1] Now if ever there were a need to step up in a disaster, the COVID-19 pandemic magnified it. It is time for Africa and its diaspora to recognize that ignorance leads to distrust and distrust leads to disunity, and all the ills that impair upward mobility, development, and sustainability. Despite the critique, there is hope in the fact that there is potential for greater production in technology and pharmaceutical development. Where there is knowledge capacity, there is hope for sustainable development. Rethinking the integration of indigenous knowledge and meaningful learning in the academy (see Dei 2000) would go a long way to encouraging the efficacy required by marginalized communities toward the creation of a sustainable philosophy of development.

Contemporary Economic Considerations

The COVID pandemic has taught us that catastrophe comes in all shapes and sizes, as does its individual impact on countries of all shapes and sizes. How societies manage disaster has much to do with whose experience it is, where that experience is located, and what resources are available to process it successfully. The case studies in this volume underscore what approaches to sustainable post-crisis recovery hold viable potential for truly equitable development.

Equity remains at the heart of this study. Over the last two decades, many emerging economies have witnessed a movement of change from authoritarian to democratic rule. Yet, the world has not equally experienced the performance of democratic theory as espoused by the most developed nations of the world. Thus, when democratic governance fails to achieve the democratic goals set by the West, crises that are exogenous to leadership in the diaspora lay bare the inadequacies of modernization theories (Gwynne 2009) and the potentially harmful consequences of relying, for example, on Keynesian interventionist perspectives (Sarwat, Mahmud, and Papageorgiou 2014) considered to be a related strategy in the philosophy of economic liberalism. Keynes's attempt to understand how spending affected output, employment, and inflation ventured a solution that governments should step in to increase demand by lowering taxes. While this could enhance economic performance, it had the potential to slow development in vulnerable communities. Resorting to Keynesian advocacy of interventionist management by governments caused many developing societies to reject governance strategies where unemployment and increased recession put leaders at risk of being seen as undemocratic and unresponsive to public economic demands. The perception of economic dissatisfaction not only led to the rejection of democratic values and political unrest but also to coups d'état at the domestic level and categorization by critics of struggling societies as failed states. As challengers to Keynesian theory acknowledge (see Friedman 2002; Lucas 2003), while there may have been a kernel of evidence supportive of Keynes's economic theory, the reality is that economic collapse in most states discredits the idea of a self-adjusting economy and fails to respect the Keynesian approach. Thus, developing economies, attempting to follow such conditions for growth find it difficult to achieve economic development as Keynesian philosophy purports.

This volume therefore challenges the assertion that faithfully following prescriptions of the West is a meaningful pathway to sustainable development in emerging democracies and offers, as a plausible alternative, the production of knowledge capital for solutions of recovery and resilience in beleaguered communities. In the wake of physical, cultural, and sociopolitical disasters across and within the diaspora, observers have recognized the need for solutions to struggle in nations like Rwanda, currently in pursuit of recovery and sustainability following the genocide there in 1994 (Ngirente 2020). As Prime Minister Ngirente shared in his keynote remarks on the vulnerabilities of the country, Rwanda's recovery agenda, following its most recent disaster, COVID-19, will rest on three pillars: "Resilience, Partnership, and Innovation."

Similar challenges to development have also been observed in Haiti, whose large population of African-descended people are still seeking recovery from the trauma of the 2010 earthquake that ravaged its territorial landscape. And, again in 2021, when disrupted by both another earthquake and ongoing negative climate effects, Haitians are still without a formula for the recovery Haiti desperately needs. These challenges are further exacerbated by the threat of global climate change. In addition, a catastrophe of leadership and failed attempts to shepherd the country out of the political and economic morass it finds itself in, Haitians cry out for new innovative approaches to recovery, seeking to generate solutions from within the country and across its own diaspora, rather than rely on the long-absent support of Western governments.

The case of Louisiana, where 33 percent of the state's population are of African descent and 58 percent may be found in its most well-known city, New Orleans (The Data Center 2022), flails in its frantic search to regain what was lost from the ravages of Hurricane Katrina in 2005 and, more recently, Hurricane Ida in 2021. New Orleans's experience with disaster is also instructive of adaptations necessary for mitigating climate change and for modeling knowledge production. The latter state still envisions restorative strategies for warding off the continued onslaught of disaster, accelerated by the incidence of climate change, while identifying new strategies for keeping pending collapse at bay.

Residents of these three diasporic locations recognize that depending on the goodwill of sympathetic benefactors to provide solutions to their developmental issues provides no glimmer of hope within the anarchical world we live. For, although theories of justice, liberalism, and realism have been generous in their promises of solution and the achievements of globalization have offered a specter of hope for recovery by way of self-help and participation in the global marketplace of ideas, development has not only tried crawling into being, but in some areas, it has collapsed under the effort of attempting to keep up with the pace of development across the globe.

Temporal Shift

The time has thus come to review and reinterpret what it means to recover and find new pathways toward development. We examine the growth of these three societies by engaging a framework for development and offering suggestions for how indigenous, cultural knowledge may be produced to yield the kind of development that does not emanate from efforts to catch up with the rest of the world that has already far-outpaced them with technological expertise. Rather, we imagine an equitable world where environmental sustainability is not wholly the purview of the wealthy in the developed North. We highlight growth in challenged, disaster-prone, diasporic communities and encourage reliance on cultural wisdoms that resonate better with traditional logic and practice and empower communities to create their own blueprints for success. In other words, we recognize that if each society examines its cultural roots, it is more likely to find embedded therein potential solutions for recovery. The key lies in the states' location of unique knowledge capital and the innovative ways in which said knowledge capital may be produced and applied sustainably.

In this volume, interdisciplinary scholars, with focus on political, psychological, anthropological, economic, and communication skills, review the impact of disaster on development and explore innovative avenues for redesigning a trajectory of growth for the sustainability of communities within the African diaspora. The editor of this volume and its contributors have extensively researched and collected data on the theme of knowledge economics. We note that developing nations, vulnerable to climate change and its environmental impacts, have had difficulty appreciating, let alone making, demands about a healthy environment because their governments have been unable to address higher-order needs of society in the face of basic survival needs. Systems theory (Easton 1965), offered as a model for advancement, has not contributed in the developing nations to the system successes achieved in the industrialized world. It is therefore helpful for a book such as this to highlight how systems of oppression across the African diaspora have been able to stagnate growth. Thus, in rediscovering the utility of collaborative, community-based efforts to produce knowledge resources aimed at recovering what was lost after encounters with underdevelopment and devastating natural and man-made disasters, innovative contributions to development can be addressed. The impact of political, economic, social, cultural, and their intersectional relationships reinforce the need for alternative solutions to issues of recovery and a quest for sustainable development, primarily after catastrophic fissures in diasporic societies.

In laying out a theoretical, conceptual framework for diasporic economies to substitute their imported, tech-based, struggling economic, politically weak, socially disorganized, and culturally confused structures with a community-based collaborative model of growth that could revitalize their societies with an innovative knowledge-sharing approach to development and environmental sustainability, we hope to identify, through a case study approach, viable strategies that may be emulated by those who are ready to look to the diaspora for guidance rather than continue to seek solutions where there are none.

Case Study Approach

Water management is perhaps one of the most immediate challenges facing countries in the Caribbean and the Americas. However, because there is little flow of information between the wealthier and poorer states of the globe, few consider the potential impact of locating problems similar to theirs and learning from those experiences. What Louisiana has learned from the resiliency of its people in New Orleans, Baton Rouge, and other parishes where indigenous communities reside, is that development can be attained through collaboration with communities across the Gulf and the knowledge of how it has been attained can then be shared with locales in the Global South where communities do not have opportunities for *knowing* to the same extent as their diasporic brethren living and growing in the developed North. Activists working on the frontline of disaster recovery have made amazing progress in building on the resilient spirits of their community elders and non-profits in the region. The challenges they have overcome, albeit located in the developed world, can easily resonate with attentive organizers in the Afro-Caribbean and Afro-Latin communities and start a much-needed movement toward greening and recovery of post-disaster communities. Post-disaster recovery in Puerto Rico, Haiti, Honduras, St. Maarten, and other diasporic spaces in the Caribbean can benefit from Louisiana's innovative leadership in relation to development and environmental sustainability.

The benefit of a case study that examines collaborative energies and post-disaster solutions for development in Louisiana, often seen as the northernmost area of the Caribbean, given similarities in its cultural make-up, should be shared with a country like Haiti, whose spatial, cultural, and social distance is not insurmountable by any imagination.

In just over a decade since disaster struck Haiti, researchers have emphasized the ways the media might contribute to connecting Haiti with new strategies for survival in the aftermath of the earthquake. The media serves as a public sphere and a civic institution that generates, disseminates, and curates a local knowledge economy in search of equitable post-disaster recovery that can benefit all Haitians. Further, it offers perspectives on the ways knowledge economies can be developed through public enterprise to include all citizens, not just those in power or the social and economic elites. Haiti and Louisiana are linked by history, food, and culture. In recent years, they are also linked by climate and the changes attending it. Louisiana has learned from Haiti in its early years of independence from France. It would do well for Haiti, given the mosaic of its past century, to learn from the progress Louisiana has made and the resilience it has demonstrated in managing and adapting to the vagaries of the ecosystem. In addition, connecting with its own large diaspora through media and creative art can help Haitians reimagine a future where they are empowered and their knowledge base advocates for the intellectual capacity lying dormant within their local communities. Through knowledge sharing, Haiti stands a chance of navigating safely through the murky economic issues, social upheaval, political illegitimacy, and the myriad other ills that plague the Global South.

Wrestling knowedge capital from the past and utilizing transdisciplinary perspectives and intersectional analysis, Rwanda's recovery plan has included ideas of justice as manifested in homegrown practices such as Gacaca, which is the hearing of local trials of persons violating community and cultural norms. These are brought before a communal committee, as in historical Rwandan tradition, to face accusers and manage conflict in a way that the community finds just. Such methodologies have been aided by education and training in academic and social institutions to repair the damage of identity wars preceding the catastrophe of 1994. Recent generations specializing in anthropology, psychology, public health, social work, and related behavioral scholarship encourage policy decisions centering women, for example, and their issues in a sociocultural environment where inequities are addressed, and solutions found, internally. As anthropologists interpret traditional practices, psychologists identify innovative formulas for relieving stress and reducing trauma, activities such as laughter yoga, dance, and handcraft have been designed to work at the community level to engage necessary conversations and behaviors that augur well for peaceful communication and Ubuntu (radical hospitality) within neighborhoods. The case study approach engaged here illustrates the impact of these methodologies and confirms the argument that knowledge capital when appropriately produced can and will contribute to sustainable development in post-disaster societies. As states within the diaspora engage in knowledge sharing and learn from each other, we envisage what in diasporic parlance is explained as "hand wash hand makes hands clean."

Development of the most pertinent goals of the UN's seventeen recommendations for sustainable development can be built on the wisdoms of the past to restore a diaspora struggling under the weight of injustice to achieve viable development. The impact of climate change is felt across the globe, varying among regions, generations, age, class, income groups, and gender. Based on the adaptations recommended by the Intergovernmental Panel on Climate Change (IPCC), it is clear that people who are already very vulnerable and most marginalized are disproportionately impacted and need innovative strategies to prevail. The poor, primarily in developing societies, and further burdened by catastrophic weather, are more likely to be in the greatest need of new practices to adapt to climate variability and change. Communities in the diaspora, whether located in the Global North or South, many struggling with coastal erosion and infrastructural failure, experience the injustices of disaster management and recognize the physical, social, cultural, and political vulnerabilities that attend them.

The impact of post-disaster experiences on Black lives complicates the struggle for development. With the evolution of social movements and the realization that the Black experience is one of underdevelopment, subpar growth, and a record of low performance across the spaces inhabited by Black residents, there is no more pertinent time than now to comprehend the implications of injustice and the barriers to development as a global disaster rooted in the cultural history of the diaspora and requiring a *Sankofa* model for change. Tackling seventeen goals for sustainable development laid out by the United Nations will help us imagine an ecosystem and a humanity that is equitably shared by those who inhabit the earth. By paying specific attention to ending poverty in all its forms and addressing economic security, which includes justice in the distribution and security of food, water, and income enough to overtop the poverty line, regions in the diaspora can begin to reclaim what was violently taken from them through slavery and the displacement of their lives across the globe. In addition, consideration must be given to the achievement of gender equality and empowerment of all women and girls, so that they may reclaim their humanity from the wanton abuse and violence young women experience via human trafficking and the slate of ills accompanying it. Theories of knowledge propose that access to education when widely available prepares publics for self-discovery, leadership, and most importantly, self-fulfillment. What locations in the diaspora have done successfully, for example, Rwanda, in allocating space for women to thrive as parliamentarians and create policy avenues for other women to chart their own development, is precisely what the diaspora can share in order for each area to design a pathway through innovative exploration to its sustainable development.

Similarly, our research establishes the view that availability and the sustainable management of water and sanitation for all ensures humanity thrives and regenerates, building on shared experiences for the sustainability of life. Addressing human vulnerabilities across communities will not only allow humans to adapt their behavior to protect the environment from a changing climate but will allow them to apply creative wisdoms for sustainability. Sharing knowledge and expertise in traditional settings helps societies in the diaspora to meet the public health standards identified by the United Nations to claim citizenship in the global commons. And, it is also important that members of the Black diaspora understand the urgency of taking action to combat climate change and its impacts. Those impacts have been severely felt in the pre-disaster diaspora. Louisiana has seen its coastline erode and its interior degrade, affecting its livelihood from marine cultivation and its tourist appeal. Its disaster experience has been expanded by the latest catastrophe to impact its way of life. The COVID-19 pandemic has limited its social activities, crippled its nightlife and the hospitality services that attended it, and severely impacted the pocket books of all those relying on the culture of the state for new expression. The harm to its built environment has laid bare the poverty that exists behind the hustle and bustle of everyday life. A loss of life, employment, and income in periods of weather disaster only expands the devastation that Louisiana communities experience. New perceived threat for the annual hurricane season, where the damage to already vulnerable communities is magnified, is perceived to increase displacement as communities repair and recover. So, where would resilience come from, if not from learned, cultural wisdoms of the past? And why should such resilience, when achieved, not be shared with other diasporic communities to empower their engagement of innovative methodologies of adaptation while advocating for new knowledge-producing mentalities.

The experiences of Haiti and Rwanda (the latter may not have a hurricane season to cause national stress but understand the trauma of other stressors) must continue to address the crises of political leadership, challenges to legitimate governance, and the juggling of management criteria to ensure nations do not regress into complacency in the face of disasters that threaten employment, income, wellness, and community, all of which are integral to recovery.

Academic institutions, especially historically black colleges and universities (HBCU), have long suffered from lack of access to strategic knowledge on which new ideas may be built. Knowledge capital, as an essential path to knowledge production, can be located in stories of the past and remedies long dismissed as old-fashioned, non-scientific and, often considered useless household practices that are behind the times. Yet, in the face of failed attempts to keep up with the pace of change and engage world-class technologies that have mired diasporic societies in debt and poverty, it is important to reclaim wisdoms of the past that are still instructive for both students and faculty. The application of knowledge capital and the production of said capital can arouse interest in innovative projects that may be shared widely within, and even beyond, the diaspora, in pursuit of sustainable and developed communities to empower and advocate for them in the management of sustainable post-disaster societies. Publication of innovative work conducted by faculty and students in response to the discussion in this text will further advance our attempt to awaken a spirit of solidarity among diasporic peoples and a will to build an edifice of knowledge in which access to collaborative knowledge sharing may be forthcoming. In the spirit of Wangari Maathai (2003), a Kenyan social, environmental, and political activist and the first African woman to win the Nobel Peace Prize (2004), the diaspora must mobilize as a unified indigenous diasporic grassroots organization that empowers its communities to address imperatives for post-disaster sustainability and resilience. And in tribute to Jewel L. Prestage, the first African American woman to receive a doctorate in political science in the United States, I echo the belief that education and knowledge sharing are the keys to development and sustainability of the African diaspora.

Pamela Waldron-Moore, the Leslie R. Jacobs Professor of liberal arts education at Xavier University of Louisiana, holds a PhD in political science with specialization in Comparative Politics and International Relations. She has taught at universities in the Caribbean and the United States. Among her research interests are the political economy of development, gendered disparity, and climate justice. She is published in peer-reviewed journals of comparative politics and international studies and contributes regularly to scholarly forums on transformative pedagogy. She is a recipient of Keynote Speaker awards for research on women and the 2018 Jewel Prestage Mentorship Award from the National Conference of Black Political Scientists (NCOBPS).

Note

1. https://www.ifc.org/wps/wcm/connect/news_ext_content/ifc_external_corporate_site/news+and+events/news/cm-stories/africa-pharma-manufacturing-hubs-en.

References

Ausubel, David. 1978. *The Psychology of Meaningful Verbal Learning.* New York: Grune and Sutton.

Dass, Sujan Kumar. 2020. *Black People Invented Everything: The Deep History of Indigenous Creativity.* Atlanta, GA: Supreme Design Publishing.

Dei, G. J. S. 2000. "Rethinking the Role of Indigenous Knowledge in the Academy." *International Journal of Inclusive Education* 4(2): 111–32.

DuBois, W. E. B. 1905. "Niagara Movement Speech." *Teaching American History.* Retrieved 18 April 2023 from https://teachingamericanhistory.org/document/niagara-movement-speech/.

Easton, David. 1965. *A Systems Analysis of Political Life.* New York: John Wiley.

Friedman, Milton. 2002. *Capitalism and Freedom.* Chicago: University of Chicago Press.

Gallagher, Kelly. 2022. "The Coming Carbon Tsunami: Developing Countries Need a New Growth Model—Before It's Too Late." *Foreign Affairs This Week,* 14 December. Retrieved 18 April 2023 from https://www.foreignaffairs.com/articles/world/2021-12-14/coming-carbon-tsunami.

Goduka, I. N. 1999. "Indigenous Epistemologies—Ways of Knowing: Affirming Our Legacy." *South African Journal of Higher Education,* 13: 26–35.

Gwynne, Robert N. 2009. "Modernization Theory." In *International Encyclopedia of Human Geography,* 2nd ed., ed. Rob Kitchin and Nigel Thrift. Amsterdam: Elsevier.

Lucas, Robert. 2003. *Macroeconomic Priorities.* Princeton, NJ: Princeton University Press.

Lumumba, P. L. O. 2022. "Africa in the Next 25 Years will be Recolonized." YouTube video, 54:56. https://www.youtube.com/watch?v=pmXSi6_vFAI&t=1287s.

Maathai, Wangari. 2003. *The Green Belt Movement: Sharing the Approach and the Experience.* New York: Lantern Books

Mpofu, E., Ntinda, K., & Oakland, T. 2012. Understanding Human Abilities in Sub-Saharan African Settings. Online Readings in Psychology and Culture, 4 (3). https://doi.org/10.9707/2307-0919.1036.

Ngara, Constantine. 2007. "African Ways of Knowing and Pedagogy Revisited." *Journal of Contemporary Issues in Education* 2(2): 7–20. https://doi.org/10.20355/C5301M.

Ngirente, Édouard. 2020. "Keynote Remarks: 2020 Concordia Africa Initiative." *Concordia,* 9 December. Retrieved 18 April 2023 from https://www.concordia.net/africa/2020digital/report/keynote-remarks-prime-minister-edouard-ngirente/.

Obotetukudo, Sol. 2001. "The African Philosophy of Development: When Localism and Traditionalism Collide with Globalism, Is 'Tele' Communication the Answer?" *Journal of Sustainable Development in Africa* 3 (2): 39–57. Retrieved 18 April 2023 from https://jsd-africa.com/Jsda/Fallwinter2001/articlespdf/ARC%20-%20African%20Philsophy%20of%20Development%20Communication.pdf.

Park, Jong H. 2002. "The East Asian Model of Economic Development and Developing Countries." *Journal of Developing Societies* 18(4): 330–53.

Rosenstein, Justin. 2012. "The Team Brain: Beyond Email, Meetings, and Middle Management." *TechCrunch+,* 17 August. Retrieved 18 April from https://techcrunch.com/2012/08/17/team-brain/.

Sarwat, Jahan, Ahmed Saber Mahmud, and Chris Papageorgiou. 2014. "What Is Keynesian Economics?" *Finance Development* 51 (3). Retrieved 18 April 2023 from https://www.imf.org/external/pubs/ft/fandd/2014/09/basics.htm.

UN. 2020. "The 17 Goals." United Nations Department of Economic and Social Affairs, Sustainable Development. Retrieved 18 April 2023 from https://sdgs.un.org/goals.

World Population Review. 2023. "Louisiana Population 2023." Louisiana Population 2023 (Demographics, Maps, Graphs). Accessed April 18, 2023. https://worldpopulationreview.com/states/louisiana-population.

CHAPTER 2

Conceptual Framework

Knowledge Capital in the African Diaspora

PAMELA WALDRON-MOORE

Political and economic development and its likely sustainability within the African diaspora has been critiqued by boundless scholars and activists since the dawn of independence in the 1960s. Reverberating throughout the 1960s was the activism of young intellectuals charging the imperialist system with sole responsibility for the economic retardation of Africa (and its diaspora) via its depletion of African wealth in order to subsidize European advancement (Rodney 1972, 1981). When European exploitation was joined and/or replaced by US capitalists, further underdevelopment of Africa and its diaspora was undertaken. Over the course of political independence granted to colonies in Africa, the Caribbean, and areas of the Americas, the spoils from Africa and its diaspora allowed the more advanced North to seek newer pastures while the exploited South frantically tried to recover lost capital and catch up with the North. Liberal and neoliberal theories with prescriptions for developing nations were found wanting, amplifying the chasm in development rather than providing sustainable solutions for reversing limitations to growth for victims of underdevelopment.

If this were a discussion about history, we would recount numerous tales of how some regions/peoples declined political independence in search of survival within protectorates and similar affiliations. Today, a number of post-colonial societies are driven to near mendicancy, as a result of the exploitation of their raw materials, on the one hand, and, on the other hand, the need to purchase manufactured products from the industrialized core in order to ensure the survival of their citizens. But, although much of this discussion is centered on the struggles of the past, it does not reside fully in the past. This chapter presents an attempt to review the intellectual capital still resident in Africa and its diaspora and to identify pathways for reproducing this capital into development options that offer room for growth and sustainability. So where does the diaspora turn to escape the stranglehold of subordination imposed by imperialists over centuries? Is there any shred of African productive capital remaining, ripe for production and rebuilding? Has everything been taken from Africa and its diaspora, including cultural values, or can remnants of its ancient civilization still be identified and catalyzed into growth options? History teaches us and Black feminists affirm that to reclaim knowledge of the past and not repeat errors of the past, we must find uniquely creative tools to dismantle the master's house and its hold on the present. This study looks to knowledge past to determine how intellectual capital might be reclaimed, produced, and shared among African-descended peoples so that they might etch out new pathways to development.

The problem is that for too long regions in the diaspora have doubted their ability and self-confidence to redeem their capacity to thrive. Elsewhere, failure to prosper has been widely blamed on the diaspora's lack of resourcefulness or on the failed leadership of their governments. This chapter will show that with innovative calibration of knowledge and its production, the diaspora can devise a trajectory for sustainable development that disrupts neoliberal ideas.

In the post-colonial era, the diaspora has encountered its share of disaster: disaster from climate; disaster from political action; disaster from economic loss, and more. In this chapter, we will address the complications for sovereignty stemming from disaster and conceptualize pathways to recovery. We will present a theoretical framework of knowledge economies in post-disaster societies that is laid out through a trans-diasporic analysis of crises in select states and localities that are part of the Black diaspora as an alternative solution to post-disaster crises. The chapter will interrogate meaningful ways to progress in a field where the high-tech-based solutions of the Global North have continuously eluded the Global South, and even some Black localities in the Global North, when it comes to post-disaster sustainability. Taking a multidisciplinary and interdisciplinary look at the production of sustainable knowledge economies growing from the ground up, including scholarship in a cross section of fields, this study aims to recenter and reframe how we approach, design, and execute sustainable development in post-disaster societies. The concept elevates knowledge found from within post-disaster societies as providing a blueprint for both a viable and equitable recovery in vulnerable spaces that have experienced both natural and manmade crises. Lastly, this chapter strives to advocate for racial justice through sustainable development as societies across the Black diaspora continue to disproportionately recover from crises.

Theoretical Overview

Omitted from a general discourse on development in US educational institutions has been the role individuals/the masses can and do play in policy-making for sustainable development. The integration of national economies via the process of globalization has advanced the notion that growth can only be achieved and sustained through the application of high-tech and manufacturing expertise. Global North countries have exemplified the adequacy of this theory via extraction of natural resources from their own as well as Global South territories, only to convert these resources with technology into manufactured goods. Theories of economic liberalism and its reported achievements over the past seventy years have exacerbated the chasms in spatial development, contributing to a lag in wealth between the Global North and the Global South (Deaton 2016; Milanovic 2016; Temin 2017). Through this prism of global development, however, advancement has been seen as singularly economic. The reality is that national development is not unidimensional. Thus, emerging nations in the Black diaspora need to identify other integral aspects of growth, as may be evident in intersections of social, cultural, political, and economic factors.

My research argues that when these intersections are taken into consideration, there is a strong motivation to push beyond prescribed Western strategies for tech-based economic development. Smaller societies, especially those in the African diaspora, would do well to consider place-based development options on a scale relevant to their own cultural spheres of reference. Rejecting sole reliance on the tech-based prescriptions of the already advanced economies in favor of new areas of production as a plausible supplement to solutions for development, can catalyze a return to cultural wisdoms and development of innovative ideas rooted in untapped sources of indigenous capital. A framework for such growth is to embrace intellectual capital and knowledge production and, by generating an innovative approach to producing said capital into strategies that meet the need for social, cultural, political, and economic decision-making, a sustainable developmental structure can emerge. The framework, perceived to be meaningful here, introduces pathways to development that begin with homegrown community ideas that, when empowered and supported through advocacy of the community's entrepreneurial value, will lead to knowledge production that may be improved over time by the feedback that comes from continuous generation and tweaking of ideas, culminating in a system of knowledge management that has the potential to sustain development for future generations (see Figure 2.1).

Figure 2.1. Essential Pathways to Knowledge Economies, Conceptual Framework. © Pamela Waldron-Moore 2023.

Theoretical Scope

Identifying the focus of interest for this study as grounded in the literature of disaster, development, regional location, and historical versus contemporary juxtaposition of time and space, it seems appropriate to narrow the discourse to post-disaster recovery in light of the challenges of climate change. Greater is the need to scrutinize development pathways after the ravages of disaster than to recount all the problems of development applicable to political and economic independence and options for self-determination in the Black diaspora. Thus, the theories discussed here will relate to democratic governance, disaster management, risk perception, and equity as they relate to knowledge capital and its production. The literature will also highlight the challenge to development of the gender gap, economic insecurity, and environmental injustice in the regions under review. In addition, steps needed for mitigating the risk to development will include a focus on community empowerment, innovative knowledge sharing and advocacy of ideas, as well as human resilience. These will illustrate the engines of adaptation that can only lead, with community action, to sustainable levels of development in post-disaster communities.

Disaster research has been studied from a variety of angles, often taking a historical approach or contemporaneous reporting as new crises occur. Seldom is there targeted examination of the implications of disaster for developing societies. In the wake of disaster, created by natural hazards or political events, there is need to scrutinize the social, political, economic, and cultural structures that have been impacted, inhibit development, and stagnate the advancement of self-determination in vulnerable societies.

Thus, this study interrogates the commitment in disaster-prone territories in the African diaspora to determining supplemental strategies that may be embraced to effect policy changes that might contribute to sustainable communities. Three societies, similar yet different, that fall into this line of investigation and are explored in this study are Louisiana, Haiti, and Rwanda. Louisiana and Haiti have both experienced physical and political disasters that have hindered, in critical ways, progress toward sustainable development. Rwanda has experienced political/cultural disaster but is moving steadily toward recovery by way of harnessing knowledge practices that stem from anthropological considerations, with the promise of achieving sustainable development in the long-term.

Should there be continued reliance in the Black diaspora on non-governmental organizations, whose expertise lies outside of the vulnerable community and whose interest is primarily in monetizing change rather than developing the community sustainably? Over the sixteen years since Hurricane Katrina, eleven years since the earthquake in Haiti, and twenty-seven years since the genocide of Rwanda, there is clear evidence that without the efforts of collaborative communities, reliance on external forces constitutes a zero-sum game in which sustainable development in the vulnerable communities is absent. And, at the time of writing (August 2021), yet another earthquake has struck Haiti, following sharply on political instability (March 2021) that has left Haiti a precarious political and environmental catastrophe. As Andre Paultre and Sarah Marsh (2021) report, Haitian publics are in despair over what they see as the disintegration of democracy in their disaster-ridden state, as well as their powerlessness to contribute to the kind of change needed to move the needle of development forward at home. Similarly in Louisiana, another devastating Category 4 hurricane (29 August 2021) has overtopped levees in areas only marginally affected by Hurricane Katrina (29 August 2005), no doubt the impact of climate change and consequences of insufficient progress on climate development over the sixteen years between Hurricanes Ida and Katrina. How much more must they do to secure the kind of developmental outcomes necessary for their survival is a question that they must both be asking without an inkling of where clear answers will arise.

As the variation in the above selected disaster locations suggest, there are common lessons of the past to be addressed prior to a search for solutions for the future. Equity is at the heart of this study. Over the last two decades, many emerging economies have witnessed a movement of change from authoritarian to democratic rule. Yet, the world has not equally experienced the performance of democratic theory as espoused by the most developed nations of the world. Thus, when democratic governance is attempted in developing nations and fails to achieve the preconditions of modernization that the literature has indicated necessary if development is to be achieved, diasporic peoples question the value of the preconditions and blame leadership in the Black diaspora for following the North's lead without exposing the inadequacies of modernization theories so many scholars seem to hold dear (e.g., Bernstein 2007; Jaquette 2018, 1982; Lewellen 2006). Along with the idea of democracy as the best path of governance for the diaspora, related gauges for when and how governments might aid the process of economic development in newly democratic societies, are found in applications of Keynesian interventions (see Eichner and Kregel 1975) to mitigate the stressors underpinning governmental management in times of economic collapse. Keynes advocated for increased government intervention via the lowering of taxes on individuals in order to stimulate demand and free societies from economic insecurity and global depression. While his demand-side economic theory was held in high regard in areas of the North, US presidents, including Barack Obama, found Keynesianism difficult to implement in the face of Federal Reserve policies (see Justice 2021).

Thus, if developed countries can find fault with economic efforts at modernization, one might well imagine how disingenuous it must seem to rational observers when the North blames governance in emerging societies for failure to improve development. As Karin Rosemblatt (2014) discusses in his related study, Latin American anthropologists have openly rejected modernization theories focused on economic, cultural, and psychological factors while supporting the view that dependency theories, linked with national and global models for economic change, may be a better course of action for development. Rosemblatt further found that many of the scholars/intellectuals of the twentieth century, who rejected US economic strategies that support a capitalist world system, strongly believed that science and technology offered better options for generating universal knowledge than parroting strategies of the North. Such knowledge, when culturally applied has the potential to dissuade developing societies from single-minded reliance on core industrialized societies and push them toward creating knowledge economies that can advance their economic sustainability. Africa and its diaspora can benefit from such insights. As some scholars have determined (see Tchamyou 2017), knowledge economies have the potential to impact African businesses positively. Yet, there is little available on the employment or success of such efforts in the diaspora. Instead, African and other diasporic businesses have continued to collaborate with and emulate practices in industrialized societies rather than engage innovative application of knowledge to transform their economies.

We therefore challenge the assertion that following prescriptions for development (such as preconditions for modernization) by the North for meaningful pathways to sustainable development in emerging democracies will yield the kind of sustainability that the diaspora requires. Thus, we offer, as a plausible alternative, the application of knowledge production to solutions of recovery and resilience in beleaguered communities. In the wake of physical, cultural, and sociopolitical disasters across and within the Black diaspora, scholars have observed the struggle of Rwanda, in pursuit of recovery and sustainability following the genocide there in 1994; Haiti, seeking recovery from the trauma of the 2010 (and 2021) earthquake that ravaged its territorial landscape, and subsequent failed political attempts to lead the country out of the economic morass it finds itself in; and, Louisiana, flailing in its desperate search to regain infrastructural control following the ravages of Hurricane Katrina (2005) and visible in the aftermath of Hurricane Ida (2021), while still envisioning potential strategies for warding off its susceptibility to disaster, exacerbated by the incidence of climate change.

These three societies should recognize that depending on the goodwill of sympathetic benefactors to provide solutions to their developmental issues provides no glimmer of hope within the anarchical world they live. Although theories of justice, liberalism, neo-liberalism, realism, and neo-realism have been generous in their promises of solutions and the achievements of globalization have offered the specter of hope for recovery by way of self-help and participation in the global marketplace of ideas, development is still not yet assured in the Black diaspora.

So, what does development mean for the diaspora? It is time for them to review and reinterpret what it means to recover and find new pathways toward development. Through an examination of growth in these three societies and engaging a framework for development, we offer suggestions for how indigenous, cultural knowledge may be produced and yield development that does not emanate from efforts to catch up with the rest of the world that has already far-outpaced vulnerable societies with technological expertise. This chapter hopes to highlight growth in challenged, disaster-prone communities and encourage reliance on cultural wisdoms that resonate better with traditional logic and practice and empower communities to create their own blueprints for success. In other words, if each society were to examine its cultural roots, it is more likely to find embedded therein potential solutions for recovery. The key lies in the states' knowledge resources and the innovative ways in which said knowledge may be produced and applied, sustainably. As interdisciplinary scholars review the impact of disaster on development and explore innovative avenues for redesigning a trajectory of growth for the sustainability of communities within the Black diaspora, it is likely that post-disaster societies may find the will and be empowered to accept the challenge of development that is imperative at this time. What is needed for the diaspora to move forward in the aftermath of disaster and the challenges that exacerbate underdevelopment is education and research, access to information and technology, gender inclusion and equity, food and economic security, and environmental and climate justice that engender collaboration at the local, national and global levels.

Efficacy of Democracy for Equitable Development

In the Black diaspora, over the last sixty years of independence, postcolonial societies, particularly in the Anglophone and Francophone regions, have engaged fully the prescription of democracy and the dividends it offers for development. Political theory is replete with a recognition of the rights of individuals, wherever they may be located. Following from John Locke's treatise on the law of nature (1690) and the more historically recent Universal Declaration of Human Rights (UDHR 1948), not only has the world embraced the notion that democracy, as a political system, entitles humans to freedom and dignity as equals but also that all people without distinction of race, sex, language, religion, national or social origin, birth or other status are entitled to the right to life, liberty, and security. Yet the world, particularly the African diaspora, has known slavery, servitude, and exploitation by the very systems that uphold the United Nations Charter and the UDHR. And though the last three decades have witnessed a movement of change from authoritarian to democratic rule, the world has not equally experienced the performance of democratic theory as espoused by the most developed nations of the world.

Eminent political theorist Robert Dahl (1989) has identified what he considers to be the most significant aspect of democracy central to a critique of democracy. He asserts that a continuing responsiveness of a government to the preferences of its people relies on the ability of government to discern these preferences as well as act upon them so that the contract between citizens and governments can be maintained. Only then can true democracy be attained and what political philosophers, such as Aristotle, called the "good life" be achieved. Thus presented, democracy and democratic values have been the catalyst for a world reformation in which most of the world's nations, under the label of democracy, have sought to experience the "good life" Aristotle envisioned.

Modern political thought has, however, invested more study in pursuit of the "good life" than Aristotle perhaps even imagined. In addition to embracing liberal democratic theory as a prime component of democracy, political scholars visualized a connection between democracy and prosperity and specifically scrutinized the relationship between democracy and economic development. This became a preoccupation of political theorists arguing about the direction of the relationship between economic development and democracy (see Ansell and Samuels 2014; Collier 1999; Huntington 1993; Inglehart and Welzel 2005; Rueschemeyer, Huber, and Stephens 1992; Sanborn and Thyne 2014; Therborn 1979; Treisman 2020), with some arguing that where there is democracy, economic development follows, and others claiming that economic development precedes the struggle for rights and liberties. Either way, it is this preoccupation that led political scholars such as John Gerring (2010) to question why democratic governance in advanced industrialized societies work so well when the same is not seen in developing societies where growth is uneven and political instability ever-present. If liberalism is a political and moral philosophy espousing liberty, equality before the law, and support for individual rights, and neo-liberals hold perceptions respectful of free markets, free trade, gender equality, racial equality, internationalism, and many other freedoms, why does it not extend justice and fairness to all humans, be they in the Global North or South where democracy and economic development are ardently pursued? Why does attainment of the good life seem within the reach of some while global exploitation of less developed countries, worldwide gender inequality, environmental degradation and, most noticeably, wide gaps in economic security persist? In other words, why does Haiti, which attained independence in 1804, fail to model democratization as an off-shoot of economic development or vice versa? Also, what aspect of the debate applies to Rwanda and, to a lesser extent, Louisiana? The answer must surely be related to inequity in all its phases (Gilens 2005).

A few scholars have sought to answer the question by suggesting that attainment of the good life requires citizen competence, perhaps something resembling "the ideal democratic citizen" (Gilens 2016) whose knowledge of the issues and the ability to form interesting and logical questions may play a role in democracy by assessing how well or if their well-being has improved or declined under democratic governance. This introduces the perception that the gap in economic development may be rooted in the intellectual capital of citizens. The politically cognizant and interested citizen would supposedly be more likely to prosper if the right questions and issue preferences were shared with a political leadership responsive to the needs of the community. This is not a lesson yet learned in any of the diasporic communities under review.

As research on the political efficacy of individuals illustrates (see Craig and Maggiotto 1981; Finifter 1970; Gamson 1968; Muller 1977; Paige 1971) low trust and high efficacy in Western societies combine to make the case for citizen activism in the decision-making process of governance moot.

While some insist that citizens who trust governments to act on their behalf and are willing to state their demands confidently are likely to perceive the responsiveness from government officials that demands continued participation in the political system, there are many more who believe that governments do not care what "people like them believe" and therefore dissociate themselves from the needs of society or engage in unconventional behaviors aimed at disrupting allegiant actions. This raises the question for both developed and developing nations about how to ensure government responsiveness and therefore sustain democracy. In developing nations where discontent and loss of trust in legitimate government may give way to non-allegiant behavior, it is often the military that engages in unconventional action, such as military coups and other forms of take-over. Haiti is among countries with a large Black diaspora that has experienced such activism to the detriment of growth and development. Crippled dissent or responsiveness does not encourage innovative creativity for a sustainable future. And, in situations where disaster mitigation is necessary, a lack of communication and consensus essentially just kicks the can down the road until the catastrophe is unmanageable.

Introducing theories of knowledge and trust into considerations of democratic governance and connection with economic development is, thus, a useful way to assess the success of advanced democracies and failures of still-emerging nations. Modernization theorists (e.g., Huntington 1984, 1991) who were once eager to embrace the fact that education was the key to development, hopeful that the Asian model of following the cues and in the footsteps of the industrialized West, would have the same effect on other parts of the world were soon challenged by dissenting scholars (e.g., Przeworski and Limongi 1993) whose growing disenchantment with modernization theory as a model for economic growth urged reconsideration of the tenets of democracy as an indicator of economic growth. Many came to the realization that development, via education and experience, perhaps precedes modernization.

As scholars sought connections, scholarly or otherwise, between economic development and democratic governance and watched former Eastern European countries claim democracy, join the West, and begin to experience modest increases in economic development, authoritarian shifts away from liberal theories of governance also became evident. The question for political scholars, aware of the backsliding into command economies in East and Central Europe, is whether the connection between political democracy and economic prosperity is not at all causal but, in fact, spurious (see Heo and Tan 2001). This leads to a contention that the developing world may be better served by interrogating reasons for their failures and identifying new ways to reform the political systems of oppression that attend economic development. History shows that developing nations have spent much time deliberating the value of democratic principles leading to development because of their fundamental faith in the ideals of fairness, equality, and justice that would help them achieve all that democracy promised rather than a recognition that control of their own natural resources had greater potential to forge economic development and the "good life" they sought than dependence on those espousing democratic ideals. Missing from the literature, however, has been the element of democracy that has fostered slow or non-existent development. So, while it is fair to focus on the recursive relationship between democratic values and economic growth in developed nations, the performance of democracy in the developing world should not be ignored as the source of its uneven development. In developed democracies, there are monetary tools in place to regulate the relationship between democratic value and economic growth. Juxtaposing expansionary monetary policy with contractionary monetary policy has allowed developed nations to implement policies in ways that achieve economic growth (O'Connell and Schmidt 2021). Diasporic systems are seldom able to put such growth into effect, based on political decision-making and, instead, this creates disenchantment between political leaders and their preferred economic stratagems. Until such lessons are learned, developing nations must decide the nature of the political system most likely to sustain ideal citizenship for enjoyment of the "good life." They could make the choice to harness their knowledge and skills capital to effect production of innovative ideas that would yield the economic growth they seek rather than dwell on the institutional values of democracy that have rendered them unequal and politically unstable.

So where does this leave post-colonial peoples, especially those of African origin in the Caribbean and the Americas? How do they engage alternative strategies and leave habits of dependency and crises of legitimacy behind to use their intellectual capacities in more productive ways with the expectation of finding, not material wealth necessarily, but certainly an appreciation for and development of traditional wisdoms/energies that many in the diaspora are ready to reclaim on the way to realizing sustainable development. As Audrey Lorde (2018) so eloquently stated, "one cannot use the master's tools to dismantle the master's house." It is time for Africa and its diaspora to rebuild with new tools, born of their knowledge capital, and produced for long-term development. The COVID-19 pandemic (2020) has reaffirmed for less developed countries the urgency of living in an anarchical world where self-help and resilience are their best resources. It is incumbent upon developing nations not to wait to be invited to the decision forums of industrialized democracies but to start organizing and building their own collaborative forums within which the survival of developing nations can be mapped.

How the diaspora will fend for itself without accumulating debt from borrowing much while gaining little will be explored in this framework of liberalism where rights inalienable to all people have been consistently denied to people of color. Only in an awakening to how marginalized peoples have been trapped, exploited, and deprived of their humanity while ownership of their natural resources has been seized as dependency on the perpetrators prevail, can developmental change occur. It is the absence of access to these inalienable rights, the experience of democracy that claims to grant equal opportunity without the potential outcomes of equality, and the continued discrediting of knowledge as a central capacity builder for nations that warrant an exploration of theories of knowledge economies, which suggest that focus on education, and innovation is critical in national planning strategies. As Jean-Eric Aubert and Jean-Louis Reiffers pointed out in their 2003 World Bank report "countries that fail to become part of the *information revolution* risk becoming even more marginalized than those left aside in the earlier industrial revolution" (2004: xii). Applying democratic theories of liberalism to theories of knowledge development in the contemporary world order makes a logical case for why adopting a knowledge economic framework in the pursuit of development may be advantageous and timely for developing nations.

Expectations of a Knowledge Economy

The idea of knowledge economies took form with World Bank identification of its key pillars:

1. Knowledge economies are institutional structures that offer incentives for entrepreneurship and the use of intellectual capital.
2. It requires a sound education and training system that prepares individuals for skilled labor.
3. It encourages a strong innovative approach by which collaboration among educators, the private sector, and communities could develop struggling societies.
4. It requires access to information and infrastructures for technology that will further enhance development.

Institutional Economic Structures

Across the Black diaspora, regimes have tried to emulate global structures (e.g., those of the West) and institutions in order to pursue effective allocation of resources and motivate citizens to create, disseminate and apply available knowledge. Often, the economic structures to be emulated were vastly dissimilar from those in the diaspora. Knowledge economies are a collaborative system of consumption and production that are generated via the intellectual capital of communities of citizens who are able/willing to convert homegrown and scientific discoveries into applied research that may benefit the community by allowing value to be assigned to both tangible and intangible assets. In the Black diaspora, in order to survive a life of exploitation and enslavement as meted out by imperialist economies, workers slowly learned that if they pooled their intellectual capital, they could experience growth and development on a scale befitting their needs. Within the Caribbean culture, for example, communities recognized that traditional customs surrounding trade and financing in the immediate post-emancipation period could become a way forward, if managed sensibly (see Moore 1987; Moore and Johnson 2000). Whether in areas of small-scale farming and agricultural produce or in collaboration on sugar, rice, and mining production, collective discourse on the value of shared intellectual property was assessed upon the attainment of political independence.

Recognizing that economic independence could not be gained by simply emulating the commercialization processes of the West, the Caribbean Community (CARICOM), founded in 1973, established the Treaty of Chaguaramas to link the economies of Barbados, Guyana, Jamaica and Trinidad and Tobago. Later, eleven other Caribbean territories joined the organization, including Haiti, which was added as recently as 2002. This regional entity set out to investigate links between the scientific community and academic scholarship to determine alternative ways to progress. Its main purpose was to promote economic integration and cooperation among its members. Today, ties with the Group of 77 (founded in 1964 with seventy-seven members), have expanded membership to 134 members of the developing world, with a significant increase in numbers coming from the Black diaspora. Also, the Non-Aligned Movement (1961), founded by Jawaharlal Nehru (India), Josip Broz Tito (Serbia), Sukarno (Indonesia), Gamal Abdel Nasser (Egypt), Kwame Nkrumah (Ghana), expanded the mission to ensure the national independence, sovereignty, territorial integrity, and security of non-aligned nations in their resistance to imperialism, colonialism, neo-colonialism, racism, and other forms of external aggression, occupation, domination, hegemonic interference, or bloc politics. Today, the movement has 147 members, observers, and international organizations, 55 percent of which are developing nations with approximately 50 percent of those including populations of Afro-diasporic peoples. Further, the ACP (African-Caribbean-Pacific) partnership, established in 1975 along with participation in other regional alliances, provides collaborative scope for knowledge economies to develop and improve trade, financing, and community-building. At the same time, community partnerships are also developing via traditional household interactions.

Although knowledge economies are receiving greater attention in the information age in which we live, they have existed in Africa and the Black diaspora for a very long time. Within and across the African diaspora, informal economic structures existed and still do, though to a smaller extent. As may be observed in areas of the Caribbean (Guyana, Haiti, Jamaica, etc.), old financial customs (e.g., those known as *partner*, *box hand*, *sou-sou*) helped grow economic communities in which families were enabled, with the help of the community, to start businesses, contract the building of houses, pay rents, and more without the help of banks to finance such projects (also see Browne 2018). Agriculture was upgraded to commercial levels based on knowledge sharing and micro-financing trade within communities. What is today observed as in-kind contributions, where simple barter techniques are used to share goods/services within the community, may be attributed to the knowledge capital shared in the past among generations of Afro-diasporic peoples. The information age, however, decelerated growth of such processes. Many developing countries shifted to manufacturing and service-based economies, illustrated by research, technical support, consulting, using old sources of knowledge but new reliance on human expertise from afar (rather than within the community) and reliant more on new innovations and greater competition that encouraged more secrets, less openness, and less sharing for community growth.

Can such institutional economic structures survive at the national level or are they better served in small-scale community administration? As the world contracts only to expand to new heights, the Black diaspora is encouraged to think globally but act locally. There is still scope for garden varieties of innovative economic structures to survive but it will take interlinking networks of efficient economic restructuring to realize the kind of sustained growth knowledge economies imagine. And, to a great extent, focus on disaster mitigation may enhance the scale of developmental outcomes diasporic communities need to survive.

Education and Training

Thriving economies need the support of an educated and skilled workforce, who need to elevate and apply their intellectual capital to expertly use the knowledge that has already been produced and help to train others in the adaptation of important skills. Basic education is needed to provide citizens with the literacy to process information. However, such education need not follow just the basic traditional Western path of primary and secondary training applied under colonial rule but should embrace technical and higher order learning in fields where technological training, in science and engineering, may enhance development and create innovative practices. Research shows that in today's world, job relevancy and cultural awareness should be at the heart of education if sustainable development is to be the end-product. According to the World Economic Forum, a majority of businesses (over 90 percent) are expediting digitization of their work process, adopting high-tech strategies, such as artificial intelligence (AI), cloud computing, big data analysis, and so forth. Many higher-education institutions in the United States are not geared to train students in areas identified by business to meet the digital demand of the twenty-first century, let alone communities in the African diaspora. The latter, in the context of compiling knowledge on post-disaster potential for development would need to conduct extensive research on the aftermath of tragedy.

As conceived by scholars, researching knowledge from tragedy (see Harrison 2021), using the attacks on New York on 11 September 2001 as a source of education, there are at least four areas of study that warrant focus:

1. Studies of physical and mental health of survivors after a catastrophe (see Alpert, Ronell and Patell, 2011; Mijanovitch and Weitzman 2010; Tosone 2011);
2. Improving a nation's preparedness and security after a crisis (e.g., Hale and Moberg 2005; Schmitt, Eisenberg, and Rao 2007; Shklovski et al. 2010);
3. Evaluating current infrastructure and developing useful adaptations toward green sustainable infrastructure and approaches to fostering resilience (Butenuth et al. 2011; Manfre et al. 2012; Patel et al. 2018); and
4. Examination of policy consequences of disaster and making preparation for future threats (Adam and Bevan 2005; Badri et al. 2006; Fengler, Ihsan, and Kaiser 2008).

There is, however, still a strong need in the Afro-diaspora for retention of past knowledge, reflection on old narratives, shared through the ages via word of mouth when griots relayed historical truths in the form of fables or proverbs to attentive communities. Many of these were passed down from generation to generation and traveled with slavery to the New World, again to be shared with communities fighting for survival. Given the mastery of resiliency in the post-emancipation era, the Black diaspora has much to draw on when facing contemporary disasters such as the genocide in Rwanda or the destruction by earthquake in Haiti. It is difficult to imagine that none of the approaches of the past have benefit for survival in the present and can be harnessed and modified for application to the present and innovative expression in the future.

Unlike in European communities, skills and competencies of African-descended peoples are not to be found in the annals of history. Archives have repeatedly silenced/erased the capital of Black people, relaying stories of power, competence, and glory attributed to any other than those of African descent. Yet, it is in knowing the past and the intellectual capital reposed in the past that the diaspora can better learn how to be resilient and manage catastrophe, how to rebuild lives, cultures, and patterns of existence that can be adapted to today's circumstances. As described by the literary accounts of scholars, the pain experienced in encounters with "scraps of knowledge in the archive," causes one to fashion and bridge the past and the present and to "dramatize the production of nothing" (see Fuentes 2016). It is this disadvantage that leaves scholars of the African diaspora to choose the path of least resistance, the choice to depend on the educational structures in the Global North rather than delve beyond the archival sources and basic instruction provided in parochial and denominational educational institutions in post-colonial societies across the Global South. Especially in relation to catastrophe in nature and risk mitigation, it is not helpful, for example, for Louisiana or Haiti to simply follow methods practiced in the Netherlands or elsewhere in Europe to resist the ill effects of climate change and its annual impact on the disastrous weather experiences of diaspora locations in these regions. One of the lessons to be learned is in knowing that old technologies, for example, building levees for protection against rising seas, are much less reliable in the twenty-first century than nature-based protections of shorelines that become stronger over time. Stone and concrete levee structures experience more erosion and decay even with appropriate levels of monitoring than structures developed with collaboration between ecologists and engineers who recognize the benefit of planting at higher densities and engaging in maintenance and systematic monitoring that creates resilient structures (Burdick 2021).

As an old African proverb suggests, "until lions have their own storytellers, hunters will always be the heroes in their story," or otherwise stated, "until the lion learns to speak, the tales of hunt will always favor the hunter." The point is that until knowledge about Africa and the diaspora is disseminated by Africans in the diaspora and their entrepreneurial skills shared and adapted to meet development needs, sustained growth and advancement will not be achieved. In her essay "Post-Disaster Futures: Hopeful Pessimism, Imperial Ruination, and *La Futura Cuir*," Yarimar Bonilla (2020) discusses how disaster-prone spaces (e.g., Puerto Rico) could imagine and rebuild futures that deviate from a knowledge of post-colonial sovereignty to a promising post-disaster future. Recognizing the psychological impact of slavery and its intergenerational legacy of trauma on the Black diaspora (see Carten 2015), there is little wonder that huge deficits in confidence and innovative creativity could enervate knowledge production and infuse, with empowerment and advocacy, a determination to migrate from pessimism and abject frustration to a new design of freedom, self-determination, and development, devoid of imperialistic tentacles and dependency. Yet, this may be seen in some circles as the only way forward to overcome the lagging development found in post-disaster communities of the diaspora. Haiti's experience is a case in point.

Innovative Systems

The Adaptation Fund, providing funding within/among countries of the Global South aims at building resilience in periphery states to adapt to climate change. Whether through a glorified *sou-sou* system, such as those engaged in the Black diaspora and on the African continent or through regional alliances, as discussed above, developing countries can benefit from opportunities to create entrepreneurial skills that might contribute to knowledge production emanating from shared ideas, financial collaborations, and community engagement. Instead of Haiti, Rwanda, and Louisiana relying fully on external measures to restore the infrastructure needed to resist climate change and its impact on natural, cultural, and political disaster, they can provide the requisite green infrastructure, with community expertise for coastal recovery and improved water quality, among other needs.

One of the innovative practices used to a fair extent in Louisiana has stemmed from the work of the Bucket Brigade, a local group of activists who share work in the academy and within the community. The Louisiana Bucket Brigade, founded in 2000, was a response to the environmental injustice perceived as a threat to Louisiana neighborhoods having to wrestle with the pollution released into the atmosphere by the state's oil refineries, chemical plants, and petrochemical infrastructure. These industrial structures moved into poor neighborhoods on the grounds that they would elevate the economic security of said neighborhoods. Instead, they brought devastation to the health and well-being of residents. The work of this organization incentivizes communities to test air quality within their neighborhoods to ensure that they are not exposed to the harmful effects of pollution. Knowledge is power. Communities are supported, empowered when they are informed, and have the capacity to build on the information and training to which they are exposed and from which they can benefit. Technology aimed at testing air quality allows the organization to develop air sampling buckets the public can use to determine action to be taken and whom to hold accountable for the costs of poor community health (Rolfes 2010). Moreover, the Bucket Brigade, registered as a 501(c)(3) nonprofit organization, is supported by private donations from both individuals and foundations who are willing to invest in the environmental health of communities (Rolfes 2010). Anne Rolfes, having served in the Peace Corps in West Africa, returned to her home state, Louisiana, to practice some of the lessons she learned in Nigeria and Benin (see Thornton 2011) about protecting the environment. As publics become knowledgeable about community action they can take to support the nonprofit as well as extend their understanding of how self-reliance may yield better information on communities' expediting the transition from fossil fuels to clean energy, they can make demands on government institutions and private industry to visualize and effect a healthy and safe community for the current times as well as for posterity.

Production of knowledge gained from attempts at reducing pollution and dwelling in a healthier environment tends to build informed communities, healthier societies, and pollution-free neighborhoods. Further, in addition to other grassroots innovations, more of which are discussed in our chapter on Louisiana, informed residents organize for participation in a fair and equitable state where ordinary citizens prosper from their knowledge acquisitions and their capacity to generate sufficient intellectual capital to help the state make a just transition toward green infrastructure and relief from climate catastrophe. In disaster-prone Louisiana, poor communities are likely to be the most negatively impacted, a result of racial and economic disparity. According to the 2020 census and other data reports, a majority of African-descended (approximately 33 percent) residents, the second highest (second only to Mississippi) concentration in the United States compared to other ancestry groups, reside in Louisiana communities. They endure poor air quality and sanitation and experience myriad health deficiencies. Empowered with the knowledge and skill to test air quality on a given day allows local residents not only to take action to protect against ill health but also to hold health officials in their communities accountable for distress to their families and other residents. Through community actions like these, and outreach to neighboring communities, community organizations such as the Walls Project in Baton Rouge, Louisiana (Williams 2021), and other organizations on the Gulf Coast, collaborative options are engaged in successful post-disaster development. In addition, with support for women's innovative practices under the auspices of the State Department, there is greater scope for addressing climate and disaster issues and sharing knowledge on disaster preparedness and recovery than ever before. Launching *The Innovation Station*, Aubrey Paris (2021) has disseminated information and advanced knowledge capital on subjects such as shoreline resilience, food security, urban conservation, green infrastructure, landscape architecture, water management, marine protection, soil restoration, and so many other areas of knowledge that need to be produced, especially on the Gulf Coast but everywhere where the disaster impacts of climate change may be found. *The Innovation Station* serves as a great example of how local and national policy operatives can collaborate in an attempt to share knowledge and grow the success of innovative methods that may adopt and transform standard practices into more culturally relevant designs for sustainability. The more collaborative the efforts, the greater the potential for knowledge sharing and the advancement of a more sustainable ecosystem.

Other innovative practices may be found in Rwanda, where there is a strong cultural belief in healing the community. After the 1994 genocide, the idea of creating opportunities for mental health and trauma recovery introduced an idea of care, using traditional and international therapies, including representations of laughter and dance. One of the grassroots organizations established there, under the leadership of Samuel Habimana, a co-author in this volume, has contributed much to healing in the quarter century since the deadly genocide of 1994. Practicing laughter yoga at the Remera campus of the University of Rwanda's College of Medicine and Health Sciences has achieved remarkable gains for communities who grow to understand better how to mitigate the effects of stress by way of laughter. Participating groups have benefited from interaction with local residents, international visitors, and peers from neighboring communities, where they exchange ideas, share vulnerabilities, and develop easy practices to de-stress and better manage health and wellness conditions.

Community development centers in neighboring towns/villages have also incorporated dance practices handed down from the past to integrate communities in the fight for cultural harmony. Dissension in the past, which disintegrated into open anger across communities, led to one of the world's greatest sociocultural disasters. Incentivizing communities to develop strategies for harmony, peace-building, and forgiveness has empowered Rwandan communities to develop a language of common understanding that portends not just to rebuild a broken culture but to build on the knowledge of the past to create a new, vibrant future where healthy citizens advocate for programs, cultural and political, to move Rwanda into a healthy, harmonious, and just environment, where knowledge of the past and recovery from past catastrophe will generate a more healthy, safe, and environmentally sound Rwanda.

Music and dance have always been an integral part of Africa and its diaspora in the Caribbean and Americas. Traditionally, singing and dancing were a popular way not only to celebrate special events but also to teach cultural beliefs and promote social values. Where these practices were conducted, communities learned their histories, made connections across groups, and gathered problem solving skills that could be transmitted from past to present and have vibrancy for the future, if preserved. Celebratory practices were accompanied by oral histories, storytelling, recitations of struggles and how they were overcome. Singing songs from the past were used to advocate for present-day action or point out creative wisdoms that may be losing their place in younger lifestyles. Within the African diaspora, music was/is used as a revolutionary, or perhaps just disruptive technique to share information. In the Caribbean, calypsos were renowned for the messages they contained. Slinger Francisco (lived 1935–2020), Grenadian calypsonian and griot, in the African tradition, was known globally as the "calypso king of the world." His contribution to the knowledge capital of Caribbean and other diasporic audiences via the indigenous African tradition of storytelling in song chronicled the issues and philosophies impacting the lives of Africans in the diaspora in ballad form. His music challenged the colonial impact on Caribbean education and beliefs, for example, in the song "Dan is the Man" (1970). In that song, Mighty Sparrow, as he was familiarly nicknamed, laments the lessons taught in Caribbean schools and commented that the only reason he did not suffer from bad education is because he was not intelligent enough to learn to read. But he satirized that had he learned to read, he would have become a "block-headed mule." The familiar Reader, *Nelson's West Indian Readers First Primer* by J. O. Cutteridge (latest edition, 2014), was required reading for primary schools across the English-speaking Caribbean as a first reader for all students, a majority being of African descent. In more recent times, Jamaican singer, Robert Nesta Marley (lived 1945–1981), fondly known as Bob Marley, called on audiences to "emancipate yourselves from mental slavery, none but ourselves can free our minds," (originally released 1980; 2020) echoing Mighty Sparrow's disrespect for the miseducation of Black people, who were transshipped to and enslaved in the Caribbean and needed to seek redemption, to free themselves from colonial teachings.

Lessons from Louisiana and Rwanda, providing other ways of *knowing*, will also serve as a template and provide a framework, if shared, for implementing meaningful ways of addressing post-disaster recovery and other issues that plague Haiti and other areas of the diaspora. Although Haiti's experience with disaster has been mostly physical, it was certainly not entirely natural. Earthquakes and hurricanes are often described as acts of God. This could be seen as either an opportunity to inspire faith-filled approaches to development or a call to the Haitian intellectual community to devise better structural measures to a recovery program. Of course, such a call would require the collaborative efforts of the citizens, the government, the technocrats, the artisans, and the international public and private sectors, as a whole, to collaborate on a master plan for rebuilding homes, communities, think-tanks, as well as seeking governmental buy-in. There is no doubt that a strong political will, bolstered by local intellectual capital and the commitment of private sector supporters can be an engine of motivation for the accomplishment of a safer, structurally secure human society.

Haiti's earthquake (2010) was a clarion call to action for the international community, as noted in a blog by Rebecca Winthrop of the Brookings Institute (2010), as well as a reminder that disaster preparedness was/is an imperative, not just for Haiti but for developing communities everywhere that lack the financial resources and the intellectual capital for managing such disasters. What would it take for appropriate mitigation efforts to be put in place not just to safeguard developing societies but to steer off the impacts of not instituting such efforts? Whether it be the costs of migration, or that of food insecurity, or achieving any of the other goals of development, what is perhaps most required is the will to ensure that democratic values of fairness, justice, equity, and the pursuit of happiness are achieved. In an ecosphere where the single human species must survive, the ethics of sharing knowledge and caring for "the least among us," as our spiritual leaders remind us, must be embraced by the entire global community. The challenges to be met for Black communities in Louisiana, Haiti, or Rwanda, if communities are to overcome disaster effects and build sustainable futures, continue to reside in solutions to economic security, gender equity, and environmental justice. As a reminder, knowledge economies offer incentives for entrepreneurship and the use of intellectual capital. It requires a good education and training system that prepares individuals for skilled labor. It encourages a strong innovative approach by which a collaboration among educators, the private sector and communities could develop struggling societies. It insists that acquisition of these elements means ensuring access to information and infrastructures for technology that will further enhance development (Chen and Dahlman 2006; Weber 2011; World Bank Group 2020).

Access to Information

Importantly, one of the pillars of knowledge economies is its access to information and infrastructures for technology that will further enhance development. This is perhaps the area most lacking in the Black diaspora.

Not only must the diaspora engage technological development to pursue growth and recovery after a disaster but it must expand its access to information in a variety of areas and provide similar access to all its communities. The challenge to achieving success in this area is structural as well as political. It is often argued that democratic governance opens the door to political freedom and economic prosperity. Regardless of the direction of this relationship (economic development → democracy **or** democracy → economic development), the challenge is for development and good governance to work together. In Africa and its diaspora, pessimistic perceptions of how democracy will be sustained (see Lemarchand 1992) and assist in fostering economic growth still prevail (O'Neil, Fields, and Share 2021).

Access to information is largely connected with access to the internet. In the developing world, only 35 percent of populations have broadband services (see Haider, McLoughlin, and Scott 2011) as opposed to nearly 80 percent in the developed world. As shown in Figure 2.2 below, African-descended people in Louisiana fare less well than white households in access to broadband services. With access to the internet, come opportunities for smart electric grids, innovative transportation designs, and intelligent new technologies. Such smart features make it possible for democracy to work globally. The disadvantage of not being digitally and equitably connected to the world and its resources do not augur well for sustainable development in the Black diaspora. Referring to Robert Dahl's (1989) perception that what makes democracy work best is the responsiveness between governments and the governed, it makes sense that where there is communication between the public and governmental leadership, the opportunity for expressing public demands, for gaining government feedback on the demands, and allowing communication to freely flow cannot but lead to government accountability and individual efficacy and trust. Before the communication revolution, governments easily got away with a lack of transparency. In today's world, growth demands easier access to information than in the past; without access, knowledge communities cannot produce sustainable development.

There are several barriers to information access in the twenty-first century. According to Akobundu Ugah (2007), not knowing what is available or desirable or what knowledge access to information may yield is a primary impediment to development. Public awareness of impending disaster, costs of disaster, sources of funding to offset disaster, the search for shelter/safety from disaster, and so on is critical to planning for and finding solutions to the problems stemming from disaster. At the World Economic Forum, Denise Nicholson (2015), a communications librarian at the University of Witwatersrand, South Africa, compellingly argued that access to

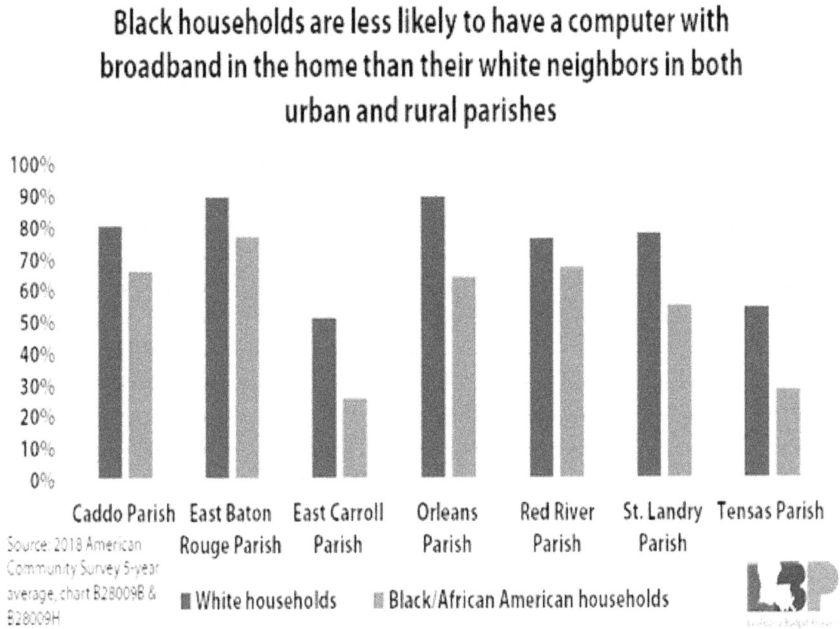

Figure 2.2. Black Households and Broadband Access in Louisiana (2018 American Community Survey included in the Louisiana Budget Project) Data taken from: https://www.labudget.org/2020/05/separate-and-unequal-students-access-to-technology-in-the-time-of-coronavirus/.

information is crucial for developing countries, if they are ever to achieve sustainable development. She challenged the government of South Africa to take a careful look at the constitution of the country and explain why that instrument acknowledged that access to information was a human right but did not take adequate steps to remove impediments to claiming those rights. Certain disaster-prone areas of the Black diaspora have been unable to benefit from necessary information because they lack continuous access to the internet. Indeed, in many countries when there is political instability, one of the first actions taken by government leaders is to cut off access to information in an effort to stifle activism and control behaviors against the lack of government responsiveness.

Other barriers, such as restrictive copyright laws, or even the absence of such laws, have a harmful impact on trade in developing countries and also contribute to economic loss. Censorship and overpricing of educational materials for libraries or for course readings and academic policies limiting publication via open access and online distribution further contribute to barriers affecting knowledge production and sociocultural development (Nicholson 2015). In a recent Louisiana Budget Project report (Butkus 2020), the state of public education as "the great equalizer" was brought into repute. With the COVID-19 pandemic moving students from classrooms to remote knowledge spaces, it was easily observed that deficits to learning were more likely to occur among low-income students of color and persons residing in rural areas than those from wealthier communities. Reflecting on the *Brown v. Board of Education* decision (1966), where the US Supreme Court outlawed segregation in public schools, political leaders were strongly encouraged to recognize the deficit in access to information and technology in public education as a form of segregation and take steps to correct the imbalance so that students of color could benefit equally from the educational resources they need to be fully informed (LBP 2020). If such deficits are observed in Black communities within the developed world, how much more likely are communities in the developing world to be barred from knowledge resources? How then might developing areas gain access to critical information and understand the urgency of managing disaster, staving off poverty, ameliorating public health concerns, and more if access to information is denied? Knowledge economies can only thrive where access to information exists and has the potential to connect communities with the past and the present to provide opportunities for future growth and development.

Of course, there are solutions for each of the known barriers to lack of information access. For the diaspora, *knowing* is critical to survival and sustainability. *Knowing* starts with access to research and research leads to innovative applications and development, which in turn leads to sustainability. Within the academy, there are opportunities to access publicly funded scholarship, which allows anyone with internet access to benefit from the research. The University of Rwanda has taken full advantage of scholarly resources on trauma and recovery and, in addition, has published the findings and innovative strategies emanating from anthropological studies, social work, psychology, and the creative arts in an attempt to build, via community action, on opportunities for recovery from the catastrophe of genocide twenty-seven years ago (Uwihangana et al. 2020).

On the other hand, Haiti has not been as fortunate in their search for and access to scientific and specialized information because their institutional affiliations have been marred by political instability and excluded from networks of organizations, such as RELX (formerly Reed Elsevier). This group serves the four largest publishing houses: Sage, Taylor & Francis, Wiley-Blackwell, and Springer. The latter has established a knowledge-sharing policy that implements limitations on what information can be accessed and by whom (see Diallo and Calland 2013). Of course, US and European academic institutions, through which Louisiana may benefit, have easy access to these scholarly resources, with abundant knowledge on a large variety of disciplines. Much of the material owned by the above-mentioned publishing houses are, however, not openly accessible (see Nicholson 2015) to all, so unless institutions in developing countries can afford to pay subscriptions for access, current knowledge and knowledge transfer activities are denied them for a period of years. While the developing world is used to barriers to growth, a barrier to knowledge, scholarship, and subsequently, advancement, in an age where information is updated rapidly, denies access that might aid development to countries that can ill afford the deficit in time and cost. Such lack of access is tantamount to cruel and degrading punishment, as may be observed in the 1966 United Nations Covenant on Civil and Political Rights, and fails to meet human rights standards of agreement as recognized by the international community (Diallo and Calland 2013; UDHR 1948).

Challenge to Development

Gendered Education

Although World Bank reports stress the importance of education to the development of knowledge economies, they seem to limit the focus to formal education, as is readily available in the developed world. In most developing countries, many people do not have the opportunity to attend formal schooling, from which research skills may arise. This lack of access to formal education hampers individuals in the developing world so that management of disasters and culling new innovative practices may not come from lessons learned in formal institutes of education but from past cultural wisdoms and in training programs observed over time between and among generations. Not including traditional, cultural knowledge as tangible knowledge capital hinders the opportunity to perceive knowledge economies as a solution for development in the Global South, especially where decisions about who gets educated is a deep cultural controversy rooted in the past. As we know, in the developing world as in a few spaces in the developed world where the wider Black diaspora is dispersed, girls are the first to be deprived of education and their intellectual capital is the first to be dismissed as uninformed. As presented by Eric Westervelt on National Public Radio's (NPR) *Morning Edition* program, systemic educational barriers have obstructed and are continuing to obstruct the progress of young Black Americans. The issue has not gone unnoticed in the policy chambers of the United States. But, former president Obama was only one of a few highly positioned leaders in past years to recognize and take steps to help close the opportunity gap for African Americans. He targeted vulnerable Black men for this initiative. Unfortunately, he did not do the same for young Black women (Westervelt 2014).

Also, since they are not engaged/included in policy analysis or sit at the decision-making tables, little to none of their intellectual capital is addressed or finds its way into the innovative policies persons in academia or government offices might need to enhance developmental structures. For such reasons, bottom-up collaborations with local communities may contribute more to knowledge production than formal lessons learned in educational institutions. Recently, the US State Department, under the guidance of Aubrey Paris, took the unusual step of hosting a series of webinars on *The Innovation Station* (see podcasts of the webinar discussions on S/GWI 2021–2023). Conversations with women and community builders, supported in these forums, have made it clear that the entrepreneurial skills of women within grassroots communities are not only of high value but offer the kind of research and data that can produce knowledge capital to meet the scale of needs in disaster-resilient communities in Louisiana and elsewhere.

Of course, in academic institutions of higher education and in government agencies where investment in research and development is abundant and where successful technological strategies have pride of place, the educated members of society must still translate the knowledge of their parents and communities into data that can then be accessed by agricultural, healthcare, and other workers whose ideas are then channeled, manufactured, and sold to help improve crops, yields, or healthcare procedures now digitally engaged for public consumption. Post-disaster recovery thus relies on the institution of cultural wisdoms within communities where they matter. In Louisiana, for example, having external specialists propose and build infrastructure for recovery of vulnerable communities often has a lower chance of success when external forces are relied on for planning and propagating new technologies for successful recovery. Following Hurricane Katrina in 2005, the Army Corps of Engineers and the national government engaged technological skills not only of the United States but of the Netherlands and other well-wishers who offered their expertise on building reinforcement for the levees that had collapsed. They quickly realized that patching the levees was not the most sustainable effort for addressing the causes of the collapse. Without consulting communities or working collaboratively with local entrepreneurs to extend and support the infrastructure to withstand climate change, repairing the levees would only provide a temporary fix for a problem that was not just technological. In both the southeast and southwest of the state, the communities' relationship with water and water management had to be addressed, providing the livelihood people needed to survive but also developing the skills that communities needed to collaborate in the rebuilding and recovery of the state. Working with the fishermen to better understand marine life and its place in the community and with farmers to care for farmlands and enrich the soil for better yields presented greater prospects for recovery than engagement in technological bandages alone.

In an interview with Aleksandra Dragozet, founder and CEO of the Sea Going Green project in the Netherlands (2021), on sustainability strategies for tourist destinations, such as New Orleans, she made it clear that publics must play a critical role when interacting with external consultancies. Dragozet was insistent that destination tourism required knowledge and innovative insights from the local community to identify areas in need of recovery and planning for sustainable development of those areas. Given her training in marine biology, she highlighted the benefit of interacting with women in the tourist community whose insights were critical to understanding the cultural-social dynamics of the impact of climate change on disaster-prone communities. Excluding community leaders, and especially the women of the communities, from the discussion table has tended to minimize opportunities for sustainable development, primarily in tourist destinations, which are hardest hit by physical disaster and climate change. Thus, the education of women, 50 percent of whom are engaged in the tourist industry in one capacity or another, is critical to sustainable tourism planning. Community outreach must therefore include education and training in relation to making the public aware of the carrying capacity of their dwelling spaces, in terms of the sewage system or impacts on the workforce in a disaster or food security in the region. So, providing women with a voice on aquaculture, or on preservation of marine life, or on conservation activities so that knowledge sharing/transfer may contribute to a transition of women's roles from lower-level engagements to decision-making at the highest level of the relevant industry, is an easy lesson for community development. Approaches taken by the Sea Going Green organization to build tailored sustainability strategies may be especially instructive to island communities in the diaspora.

Economic Insecurity

Transitions from manual to tech-based strategies are not without difficulties. Attempts to close the opportunity gap for many in the Black diaspora have been slow to yield the kind of closure that may be desirable for development. An alarming example of the impact of tech-based strategies introduced in developing countries for economic recovery is the experience of Ghana in its transition from agricultural development to new industries based on traditional agricultural yields. The process may be highlighted by the experience of Ghana in its painful transition from reliance on local agriculture to mass production of abundant agricultural resources. The cocoa pod, in Ghana, provides an insightful example of the difficulties that attend manufacturing. Converting the cocoa pod into chocolate products, because of a lack of technological expertise, was outsourced to be manufactured in the North. The low price at which the cocoa was sold within Ghana, given its abundant growth, paled into insignificance against the price Ghanaians paid to purchase the manufactured chocolate generated from the cocoa. A few missteps occurred. (1) Had Ghanaians built on the low-tech strategies adopted in the past, they may have benefited more, initially, from converting the abundant raw material (cocoa pods), easily available in the communities, into a homegrown confection which could then have inspired growth into a larger community-based industry; (2) being too swift to accept a monetary package in exchange for product without performing a competent risk-benefit analysis is a mistake that Ghana and many developing countries make in the name of short-term gains of survival; (3) innovative ideas/skills must be systematic rather than sporadic. As a recent survey suggests, 75–100 percent of innovation challenges result in an idea that can be implemented. Knowing how to evaluate success with innovative programs is part of the *knowing* that is required for knowledge economies to develop. Initially, Ghanaians may have perceived it a herculean effort to manufacture quality chocolate such as that available for purchase in Europe and North America. In their need to survive, however, they relinquished a chance for long-term gains by accepting a short-term financial boost from the sale of their cocoa pods and ultimately ended up purchasing the manufactured cocoa pods, now artfully converted into quality chocolate, at a higher price than what the unit price of the product was.

The reality, though, is that in post-colonial societies, especially within the Black diaspora, the craftiness of capitalism is pervasive. Local African-descended communities are culturally oriented toward sharing and enjoying communal traditions. They readily illustrate their prowess in the kitchen, share recipes, demonstrate traditional practices and invite friends into their celebrations. They share knowledge with outsiders without thought of patenting ideas and practices and easily bargain away their rights in the face of crafty negotiation. They seldom exercise the kind of scrutiny/analysis of the salesmanship of external producers nor do they value their own innovative cultural competencies in the art of food, spices, and other production, and the resilience of their ancestors who had successfully converted scraps from the master's table into delicious, homegrown cuisines that were amplified and enjoyed by entire communities of food innovators. So, in Africa and its diaspora, it took a while to recognize

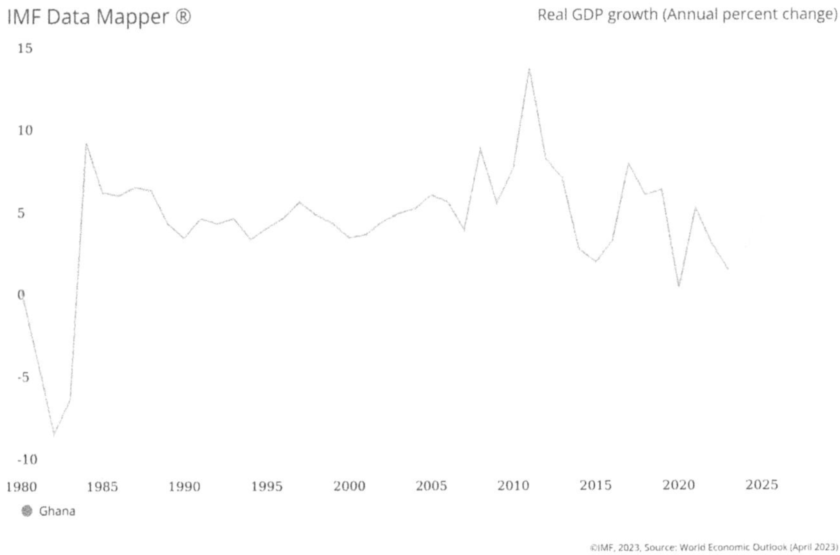

Figure 2.3. Real GDP growth in Ghana 1980-2028. Data from: https://www.imf.org/external/datamapper/NGDP_RPCH@WEO/GHA?zoom=GHA&highlight=GHA.

how their intellectual capital was being used and how their communities were purchasing goods manufactured overseas at a price higher than that gained from selling their abundant but perishable crops. By the time community leaders/elders recognized the loss of capital and resources and the benefits of developing their own chocolate, for example, and marketing it for their own profit, Europeans and Americans had already cornered the market.

Comparing economic realities of Ghana in 2001 and 2005, it was evident that Black communities were learning from their economic failures at home that success would only come with a return to collaborative norms and a deeper understanding of how to be entrepreneurs of their own capital. With the help of the World Bank, IMF, and collaborations within the private and public sector, Ghana made a significant economic turnaround. The structural adjustment program, supported by the IMF and international organizations, gave Ghana the space it needed to grow and as shown in the figure above and in World Bank data (2019), Ghana staged a remarkable comeback and is poised to become one of the world's fastest growing economies in the decade.

The economic forecast for Ghana is not unique to the development of post-colonial societies. Very similar forecasts have been made for other areas in the Black diaspora. The potential oil wealth forecast for Ghana has also been predicted in Guyana, where centuries ago, slaves were removed from Ghana, the former Gold Coast, to inhabit the highly forested and mineral rich Guianas, occupied concurrently by the British, French, and Dutch in their search for gold and precious stones. Ultimately, agreement among European colonizers divided the territory into three states—Guyana (formerly British Guiana), Suriname (formerly Dutch Guiana), and French Guiana (still under the protection of France). Just a few years ago, technology-supported strategies uncovered rich deposits of oil in Ghana and Guyana, its Black diaspora, and the scramble for wealth (this time, black gold) has resumed.

Environmental Injustice

To some extent, this turn of events contradicts the argument that the Black diaspora need not rely as heavily as they do on tech-based solutions to the problem of development as curated by the West. But, on a more realistic scale, the argument is supported that without the collaboration of educators, local enterprise, the private and public sector, development in the diaspora will not be achieved. In the geographical regions under discussion here, the obstacle to development over the years has been the rape of natural resources by capitalist conglomerates and multinational exploitation. As communities have grown and collaborations have improved, there has also been the recognition that ideas alone cannot yield development nor can economic recovery be sustained without technological intervention. Yet, this recognition does not mean that communities will/should sit back and allow themselves to be exploited. That ship has sailed.

Environmental lessons of the past, for example, Nigeria's experience with Shell (see O'Neil et al. 2021) or Ecuador's experience with Chevron (Randazzo 2021), as well as the Deepwater Horizon oil spill (April 2010) that severely impacted Louisiana wildlife, have allowed communities to find their voice and not only push political leaders to hold multinationals accountable but also engaged sister communities in their efforts to find international support from governmental and non-governmental organizations to learn the science and develop changes in regulations to stop the endangerment of offshore drilling (Luft, Korin, and Gupta 2010; Meiners 2020; Oatley 2019). While many of those dangers still remain harmful to marine life, households, and industry, the search continues for a better understanding of the science of oil drilling as well as how marine life may be protected from oil leaks and harmful toxins. In actively pursuing sustainable goals related to the UN's commitment to reversing the negative impact of global climate change on places and peoples experiencing economic insecurity, communities in the Black diaspora are challenging exploitation of their natural resources, inequity in the distribution of environmental rights, and continued governmental and external disregard for human resources vested in every community (O'Neil et al. 2021).

Other lessons of history and the past for the African diaspora teach the importance of knowledge sharing but also the wisdom of owning one's knowledge capital. The benefits of collaboration empower communities to capitalize on the scientific knowledge needed for development as well as embrace and advocate for the sociocultural traditions that may yield development. Communities also understand the value of revering traditional practices as a community, patenting ideas as a social responsibility and working collaboratively toward economic development and collective recovery in pre- and post-disaster environments. Recognizing that collaboration and sharing of ideas within communities is a more innovative way to build knowledge economies is the lesson many countries in the Black diaspora need to understand.

Sadly, this is not an area of education that communities in the Black diaspora explore. They have been denied their rights, stripped of their cultures and made to assimilate into the culture of their oppressors for so long that it is not surprising that they unknowingly teach their cultural wisdoms to imperialists and developers who, without leaders, patents, and community ownership to restrict them, capitalize on exploited communities' knowledge capital, produce said capital, and coerce or charm impacted communities into buying back their cultural wisdoms from exploitative agents. As Walter Rodney's (1972) historical writings and revisits to his work by West Indian and other scholars argue, Africa was drawn into a world economy that reduced the entire continent into underdevelopment. As the diaspora spread, taking their vulnerabilities with them, there has been full agreement with Black feminists (e.g., Lorde 2018) who recognize that the Black diaspora needs new tools to dismantle the colonial, imperialist economic and social structures that persist in disrupting pathways to sustainable development. Post-disaster communities such as Louisiana on the Gulf Coast of Mexico or Haiti in the French Caribbean or Rwanda in East-Central Africa can only disrupt the downward spiral of development by building knowledge economies partnered with the social and scientific community, public and private sectors, in bottom-up innovative arrangements built on the knowledge capital collectively produced and grounded in intersectional sharing and practices. Through the lenses of gender inequality, economic insecurity, and environmental injustice, this study can implement strategies for sustainable development.

Lessons in Innovative Practices

So, how may sites in the Black diaspora move forward innovatively toward sustainable development? Borrowing from a series of informative webinars centering community action for disaster resilience, there are many lessons on how innovations in resilience may be addressed via the lens of gender equity, economic security, and environmental justice posited in this volume as being critical foci for sustainability in post-disaster communities. In July 2021, Dr. Aubrey Paris of the US State Department, in her capacity as policy advisor for gender, climate, and innovation, highlighted the phenomenal innovations being orchestrated by women on Gulf Coast states. Titled *The Innovation Station,* Paris launched the inaugural discourse in the series, featuring women in the coastal economy, women in the advancement of food security, and alliances across these communities as they expand innovative practices for the development of local, national, and global communities. In Louisiana, with regard to engagement in innovative projects, such as shoreline resiliency, much was shared on collaborative water management, urban conservancy, and Africentric cultural interrogation to assist with artistic illustration of water management (Dandridge 2021).

Discussion of the importance of highlighting women's voices and expertise at the grassroots level (Bowie 2021) when creatively leveraging efforts in coastal resilience across the public and private sector was instructive. For example, reusing silt from dredging as part of the process of restoration is beneficial to new infrastructural design that attracts resource-sharing commitment from business as well as public partners at affordable levels. Cost sharing in terms of money, expertise, and time brings entire communities together in providing green infrastructure with the capacity to endure. Such collaborative exercises reinforce the view that knowledge production "takes a village" and when communities learn together, they grow together. Further, as persuasively argued by the Louisiana presenter (Dandridge 2021), when states invest in whole communities, and women are invited into discourse on development and groups get to listen to what the marginalized or silenced voices of women have to contribute, benefits to the state become tangible. Further, she cautioned that "building bridges without investing in people *first* is a recipe for disaster." Inclusion of women in deliberation on disaster recovery has the potential to close the gender gap, enhance workforce development, empower communities, and build resilience for a sustainable future.

Relevant to global efforts at recovery, especially in spaces like Haiti where tourism has declined since the earthquake of 2010 and economic security has become even more tenuous than prior to the disaster, the importance of information to the knowledge/education of publics is unquestionable. Political instability in Haiti, along with the lack of economic vitality there, must rely on collaboration and investment. Consistent access to the internet and social media may allow for innovations and critical rebuilding expertise to be shared. Lessons on food security, as shared by Karina Campos (2021) whose work in Argentina in combating food waste by moving agricultural surplus from the cities to rural areas experiencing drought and lacking in nutrition are quite helpful to knowledge building and production. Innovative ideas on packaging and redistributing food, creating accessibility to fruits and vegetables are supportive of public health and beneficial to the economy as it contributes to expanding the workforce as well as educating publics of efforts communities can take in collaboration with each other rather than reliance on the government. In both Haiti and Rwanda, excessive reliance on governments has led more to instability and economic deficiency than to innovative redesigning of the economic order at the local/national level.

Building knowledge economies that target development requires community access to information, advocacy for new self-help strategies, and an explosion of new, creative ideas for funding of infrastructural projects, diversification of old skills, and empowerment of communities for collaborative production/advancement. This means engaging nonprofit, for-profit, big business, political strategists, artists/artisans, and a wide variety of experts to participate in knowledge sharing, knowledge production and new vision for sustainable development in disaster-impacted communities. The impact of climate change, which has the single most devastating effect on disaster recovery, rests on adaptations for reducing reliance on yesteryear's practices of neglect resulting in food deserts, coastal erosion, destruction of marine life, inaccessibility to knowledge/education, and the devaluing of social entrepreneurship (Dragozet, Gurung, and Williams 2021).

Conclusion: Knowledge Transitions for Sustainability

Tracing the trajectory of knowledge economies over decades, we learn that the liberal international economic order (LIEO), proposed in the mid-1940s, offered a set of global, regulated, structural relationships embedded in political and economic liberalism, supported by liberal internationalism. This international order was grounded in a set of norms, rules, and institutions that recognized the bonds of friendship between European countries and the United States. Although seemingly open to inclusion of rising powers in the East, for example, China and Russia, there was not much consideration given to the interests of the less developed nations. International theorists did not lose time pointing out the restrictive nature of LIEO regulations and its hyper-globalization that undermines democracy (Ikenberry 2018; Rodrik 2012). Initially, harmony prevailed among the architects of the order. As challenges arose, however, many of the developing nations, recognizing their exclusion from the spoils of liberalism, began a search for a new international economic order (NIEO) in 1974. In the latter, developing countries advocated for an end to neo-colonialism and dependency and sought to promote South–South economic cooperation, founded on equity, sovereign equality, and elimination of the widening gap in economic and social development between the Global North and the Global South (see Marklund 2020).

In this contentious economic atmosphere, especially for Africa and its diaspora, ruminating on a framework for development exposed very harsh realities of why these regions may seem to have stagnated and, further, why there seems such little global confidence that Africa and the diaspora can catch up with the developed world and achieve a semblance of political, economic and sociocultural sustainability. As the decades since independence roll by, many ponder the prospects for Afro-diasporic communities managing well, at a time when the rise of environmentalism, the need to stem climate change, the impact of climate change on fomenting greater environmental disaster, and international advocacy for mitigation strategies by 2030, are major imperatives for the globe. This study has posited the view that a new framework for development be examined with a view to catalyzing engines of growth and reviving the potential for sustainable development in vulnerable diasporic communities such as Louisiana, Haiti, and Rwanda.

Reflecting on the World Bank Knowledge Economy Assessment pillars, it is recognized that when knowledge is the main driver of economic growth, conditions are favorable for sustainable development. What makes conditions favorable are sustained investments in education, sustained opportunities for innovation, and sustained access to information and information technologies (Chen and Dahlman 2006). First, the diaspora must overcome the barriers established by the psychological devastation of slavery; for example, lack of cultural confidence and individual and collective self-contempt; extricate itself from the political stranglehold of post-colonial dependency, and claim full sovereignty for self-determination; "emancipate yourself from mental slavery" (Marley 2020) and be innovative and imaginative in developing ideas for the future; limit investment in the global marketplace of ideas and choose to embrace new ideas aimed at overcoming disadvantages of geography and climate; and, most importantly, invest in its people by building universal education to aid research and development, universal healthcare to create better, stronger, healthier communities; enabling access to digital sources of information, and be less timid to innovate creative policies on how countries may transform marine life and agricultural inadequacy that lead to food deserts. Engaging in knowledge economies of scale portends well for the Black diaspora.

Without question, the experience of slavery has taught the Black diaspora resiliency. Such resiliency has made Afro-diasporic communities competent to manage their susceptibility to disaster, to imagine solutions for humanizing the world, and to dispense social justice to all. The four pillars of success in establishing knowledge economies are all within reach of the diaspora: institutional economic structures, on a scale relevant to the individual country; education and training in areas relevant to the needs of the community; innovative systems in pertinent areas of invention and in harmony with technology, artists, pioneers, healthcare workers, and so on; and access to digital information to update strategies, empower developers, and advocate for the health and safety of the labor force. Relying on their own cultural values and supported by their own transformative will, the Black diaspora can divine its own trajectories for development and, moreover, initiate sustainable practices for posterity. The most difficult hurdle to overcome, however, is inequity—structural and legal. If/when these are successfully disrupted, sustainable development will be achievable.

Pamela Waldron-Moore, the Leslie R. Jacobs Professor of liberal arts education at Xavier University of Louisiana, holds a PhD in political science with specialization in Comparative Politics and International Relations. She has taught at universities in the Caribbean and the United States. Among her research interests are the political economy of development, gendered disparity, and climate justice. She is published in peer-reviewed journals of comparative politics and international studies and contributes regularly to scholarly forums on transformative pedagogy. She is a recipient of Keynote Speaker awards for research on women and the 2018 Jewel Prestage Mentorship Award from the National Conference of Black Political Scientists (NCOBPS).

References

Adam, Christopher, and David Bevan. 2005. "Fiscal Deficits and Growth in Developing Countries." *Journal of Public Economics* 89(4): 571–97.

Alpert, Judith, Avital Ronell, and Shireen Patell. 2011. "Enduring September 11 in New York: Lessons Learned by the NYU Community." *Traumatology* 17(3): 3–6.

Ansell, Ben, and David Samuels. 2014. *Inequality and Democratization: An Elite-Competition Approach*. New York: Cambridge University Press.

Aubert, Jean-Eric, and Jean-Louis Reiffers. 2003. *Knowledge Economies in the Middle East and North Africa: Toward New Development Strategies*. Washington, DC: The World Bank.

Badri, Seyed, Ali Asgary, Afsoon Eftekhari, and Jason Levy. 2006. "Post-Disaster Resettlement, Development and Change: A Case Study of the 1990 Manjil Earthquake in Iran." *Disasters* 30(4): 451–68.

Bernstein, Richard. 2007. *The Restructuring of Social and Political Theory*. Philadelphia: University of Pennsylvania Press

Bonilla, Yarimar. 2020. "Post-Disaster Futures: Hopeful Pessimism, Imperial Ruination, and *La Futura Cuir*." *Small Axe: A Caribbean Journal of Criticism* 24(2): 147–62.

Browne, Murphy. 2018. "Box Hand, Pardner, Susu, SouSou." *Linkedin*, 30 October. Retrieved 21 April 2023 from https://www.linkedin.com/pulse/box-hand-pardner-susu-sousou-murphy-browne.

Burdick, David. 2021. Panel Presentation on "Green and Nature-Based Infrastructure Solutions for Economic Recovery in New England." *American Association of Atmospheric Sciences*. Webinar, October 27.

Butenuth, Matthias, Daniel Frey, Allan Nielsen, and Henning Skriver. 2011. "Infrastructure Assessment for Disaster Management Using Multi-Sensor and Multi-Temporal Remote Sensing Imagery." *International Journal of Remote Sensing* 32(23): 8575–94.

Butkus, Neva. 2020. "Separate and Unequal: Students' Access to Technology in the Time of Coronavirus." *Louisiana Budget Project*, 19 May. Retrieved 21 April 2023 from https://www.labudget.org/2020/05/separate-and-unequal-students-access-to-technology-in-the-time-of-coronavirus/.

Campos, Karina, Helena Williams, and Jeannette Gurung. 2021. "Innovations in Food Security." In *S/GWI's Innovation Station*, podcast moderated by Dr. Aubrey Paris, 15 July. Secretary's Office of Global Women's Issues, US Department of State. Retrieved 21 April 2023 from https://podcasts.apple.com/us/podcast/3-innovations-in-food-security/id1599344045?i=1000548181037.

CARICOM. 1973. "Treaty of Chaguaramas." Retrieved 21 April 2023 from https://caricom.org/documents/4906-revised_treaty-text.pdf.

Carten, Alma. 2015. "How the Legacy of Slavery Affects the Mental Health of Black Americans Today." *The Conversation*, 27 July. Retrieved 21 April 2023 from https://theconversation.com/how-the-legacy-of-slavery-affects-the-mental-health-of-black-americans-today-44642.

Chen, Derek, and Carl Dahlman. 2006. *The Knowledge Economy, The KAM Methodology, and World Bank Operations*. Washington, DC: World Bank Institute.

Collier, Paul. 1999. "On the Economic Consequences of Civil War." *Oxford Academic Journals* 51(1): 168–83.

Craig, Stephen, and Michael Maggiotto. 1982. "Measuring Political Efficacy." *Political Methodology* 8(3): 85–109.

Cutteridge, J. O. (1940) 2014. *Nelson's West Indian Readers First Primer*. Oxford: Oxford University Press.

Dahl, Robert. 1989. *Democracy and Its Critics*. New Haven, CT: Yale University Press.

Dandridge, Jessica, and Laura Bowie. 2021. "Innovations in Shoreline Resilience." In *S/GWI's Innovation Station*, podcast moderated by Dr. Aubrey Paris, 15 July. Secretary's Office of Global Women's Issues, US Department of State. Retrieved 21 April 2023 from https://podcasts.apple.com/us/podcast/1-innovations-in-shoreline-resilience/id1599344045?i=1000546837221.

Deaton, Angus. 2016. "Measuring and Understanding Behavior, Welfare, and Poverty." *American Economic Review* 106(6): 1221–43.
Diallo, Fatima, and Richard Calland. 2013. *Access to Information in Africa: Law, Culture and Practice*. Boston: Brill.
Dragozet, Aleksandra. 2021. *Sea Going Green Project*. Amsterdam: Netherlands.
Dragozet, Aleksandra, Jeanette Gurung, and Helena Williams. 2021. "Innovations in Food Accessibility." In *S/GWI's Innovation Station*, podcast moderated by Dr. Aubrey Paris, 15 July. Secretary's Office of Global Women's Issues, US Department of State. Retrieved 21 April 2023 from https://podcasts.apple.com/us/podcast/1-innovations-in-shoreline-resilience/id1599344045?i=1000546837221.
Eichner, Alfred, and Jan Kregel. 1975. "An Essay on Post-Keynesian Theory: A New Paradigm in Economics." *Journal of Economic Literature* 13(4): 1293–1314.
Fengler, Wolfgang, Ahya Ihsan, and Kai Kaiser. 2008. "Managing Post Disaster Reconstruction Finance: International Experience in Public Financial Management." *World Bank Policy Research Working Paper No. 4475*. https://doi.org/10.1596/1813-9450-4475.
Finifter, Ada. 1970. "Dimensions of Political Alienation." *American Political Science Review* 64(2): 389–410.
Francisco, Slinger. 1970. "Calypso: Dan is the Man." *YouTube*, uploaded by IsDePanInMe on 23 January 2011. Retrieved 21 April 2023 from https://www.youtube.com/watch?v=24rQOy2HZFE.
Fuentes, Marisa. 2016. *Dispossessed Lives, Enslaved Women, Violence, and the Archives*. Philadelphia: University of Pennsylvania Press.
Gamson, William. 1968. *Stable Unrepresentation in American Society*. Homewood, IL: Dorsey Press.
Gerring, John. 2010. *A Centripetal Theory of Democratic Governance*. Boston: Cambridge University Press
Gilens, Martin. 2005. "Inequality and Democratic Responsiveness." *Public Opinion Quarterly* 69(5): 778–96.
———. 2016. "Two-Thirds Full? Citizen Competence and Democratic Governance." In *New Directions in American Politics*, 2nd edn., ed. Adam Berinsky, 52–76. New York: Routledge.
Gurung, Jeanette. 2021 Innovations in Food Security." In *S/GWI's Innovation Station*, podcast moderated by Dr. Aubrey Paris, 15 July. Secretary's Office of Global Women's Issues, US Department of State. Retrieved 21 April 2023 from https://podcasts.apple.com/us/podcast/3-innovations-in-food-security/id1599344045?i=1000548181037.
Haider, Huma, Claire McLoughlin, and Zoë Scott. 2011. *Communication and Governance: Topic Guide*. Birmingham: University of Birmingham Press.
Hale, Trevor, and Christopher Moberg. 2005. "Improving Supply Chain Disaster Preparedness: A Decision Process for Secure Site Location." *International Journal of Physical Distribution & Logistics Management* 35(3): 195–207.
Harrison, Rachel. 2021. "Knowledge from Tragedy: NYU Research Post–9/11." *NYU*, 1 September.
Heo, Uk, and Alexander C. Tan. 2001. Democracy and Economic Growth: A Causal Analysis. *Comparative Politics*, 33(4): 463–73.
Huntington, Samuel. 1984. "Will More Countries Become Democratic?" *Political Science Quarterly* 99(2): 193–218.
———. 1991. "How Countries Democratize." *Political Science Quarterly* 106(4): 579–616.

———. 1993. "Clash of Civilizations." *Foreign Affairs* 72(3): 22–49.
Ikenberry, G. John. 2018. "End of Liberal International Order." *International Affairs* 94(1): 7–23.
Inglehart, Ronald, and Christian Welzel. 2005. *Modernization, Cultural Change and Democracy: The Human Development Sequence*. New York: Cambridge University Press.
Jaquette, Jane. 1982. "Women and Modernization Theory." *World Politics* 34(2): 267–84.
Jaquette, Jane, ed. 2018. *The Women's Movement in Latin America: Participation and Democracy*. New York: Routledge.
Lemarchand, René. 1992. "Africa's Troubled Transitions." *Journal of Democracy* 3(4): 98–109.
Lewellen, Katharina. 2006. "Financing Decisions When Managers Are Risk Averse." *Journal of Financial Economics* 82(3): 551–89.
Lorde, Audre. 2018. *The Master's Tools Will Never Dismantle the Master's House*. London: Penguin Classics.
Luft, Gal, Anne Korin, and Eshita Gupta. 2010. "Energy Security and Climate Change: A Tenuous Link." *The Routledge Handbook of Energy Security*, ed. Benjamin K. Sovacool, 61–73. London: Routledge.
Manfre, Luiz, Eliane Hirata, Janaína Silva, Eduardo Shinohara, Mariana Giannotti, Anna Larocca, and José Quintanilha. 2012. "An Analysis of Geospatial Technologies for Risk and Natural Disaster Management." *ISPRS International Journal of Geo-Information* 1: 166–85.
Marklund, Carl. 2020. "Double Loyalties? Small-State Solidarity and the Debates on New International Economic Order (NIEO) in Sweden during the Long 1970s." *Scandinavian Journal of History* 45(3): 384–406. https://doi.org/10.1080/03468755.2019.1624606.
Marley, Bob. 2020. "Bob Marley & The Wailers—Redemption Song." *YouTube*, uploaded by Bob Marley on 5 February. Retrieved 21 April 2023 from https://www.youtube.com/watch?v=yv5xonFSC4c.
Meiners, Joan. 2020. *Ten Years Later, BP Oil Spill Continues to Harm Wildlife—Especially Dolphins*. Washington, DC: National Geographic.
Mijanovitch, Tod, and Beth Weitzman. 2010. "Disaster in Context: The Effects of 9/11 on Youths Distant from the Attacks." *Community Mental Health Journal* 46(6): 601–11.
Milanovic, Branko. 2016. *Global Inequality: A New Approach for the Age of Globalization*. Cambridge, MA: Harvard University Press.
Moore, Brian L. 1987. *Race, Power, and Social Segmentation in Colonial Society: Guyana after Slavery, 1838–1891*. New York: Gordon & Breach Science Publishers.
Moore, Brian, and Michele Johnson, eds. 2000. *Squalid Kingston: How the Poor Lived, Moved and Had Their Being*. Kingston, JA: The Social History Project.
Muller, Edward. 1977. "Behavioral Correlates of Political Support." *American Political Science Review* 71 (June): 454–67.
Nicholson, Denise. 2015. "Why Open Access to Information is Crucial for Developing Countries." World Economic Forum, 21 August. Retrieved 21 April 2023 from https://www.weforum.org/agenda/2015/08/why-open-access-to-information-is-crucial-for-developing-countries/.
O'Neil, Patrick H., Karl Fields, and Don Share. 2021. *Cases and Concepts in Comparative Politics*. New York: W. W. Norton.
Oatley, Thomas. 2019. *International Political Economy*. London: Routledge

O'Connell, Brian. 2021. "Monetary Policy: How Central Banks Regulate The Economy." *Tucson.com*, May 14. Retrieved 21 April 2023 from https://tucson.com/business/investment/monetary-policy-how-central-banks-regulate-the-economy/article_e6f7313e-3d61-5501-8620-d83f3824dd0e.html.

Paige, Jeffery. 1971. "Political Orientation and Riot Participation." *American Sociological Review* 36(5): 810–20.

Paris, Aubrey. 2021. *The Innovation Station Series: Conversations with Women on Disaster Mitigation and Innovative Practices*. Wash, DC: US Department of State. Retrieved 21 April 2023 from https://podcasts.apple.com/us/podcast/s-gwis-innovation-station/id1599344045.

Patel, Megha, Leia Saltzman, Regardt Ferreira, and Amy Lesen. 2018. "Resilience: Examining the Impacts of the Deepwater Horizon Oil Spill on the Gulf Coast Vietnamese American Community." *Social Sciences* 7(10): 203. https://doi.org/10.3390/socsci8030074.

Paultre, Andre, and Sarah Marsh. 2021. "Quake Kills Hundreds in Haiti, Worsening Caribbean Nation's Plight." Reuters, 15 August. Retrieved 21 April 2023 from https://www.reuters.com/business/environment/magnitude-7-quake-strikes-western-haiti-usgs-2021-08-14/.

Przeworski, Adam, and Fernando Limongi. 1993. "Political Regimes and Economic Growth." *Journal of Economic Perspectives* 7(3) 51–69.

Randazzo, Sara. 2021. "Litigation Without End: Chevron Battles On in 28-Year-Old Ecuador Lawsuit." *Wall Street Journal*, 2 May. Retrieved 21 April 2023 from https://www.wsj.com/articles/litigation-without-end-chevron-battles-on-in-28-year-old-ecuador-lawsuit-11619975500.

Rodney, Walter. 1972. *How Europe Underdeveloped Africa*. London: Bogle L'Ouverture Publications

———. 1981. *How Europe Underdeveloped Africa*. Washington, DC: Howard University Press.

Rodrik, Dani. 2012. *No More Growth Miracles*. Cambridge, MA: Project Syndicate.

Rolfes, Anne. 2010. "Louisiana Bucket Brigade and Uses of Dispersants in Gulf Oil Spill." Statement before the Subcommittee of the Committee on Appropriations, United States Senate, One Hundred Eleventh Congress, Second Session, July 15, page 43, 45. https://www.appropriations.senate.gov/imo/media/doc/hearings/07_15_10%20CJS%20Dispersants%20in%20Oil%20Spill%20GPO%20Record.pdf.

Rosemblatt, Karin. 2014. "Modernization, Dependency, and the Global in Mexican Critiques of Anthropology." *Journal of Global History* 9(1): 94–121.

Rueschemeyer, Dietrich, Evelyne Huber, and John Stephens. 1992. "Capitalist Development and Democracy." *Contemporary Sociology* 72(3): 243–48.

Sanborn, Howard, and Clayton Thyne. 2014. "Learning Democracy: Education and the Fall of Authoritarian Regimes." *British Journal of Political Science* 44(4): 773–97.

Schmitt, Ted, Jon Eisenberg, and Ramesh Rao. 2007. *Improving Disaster Management: The Role of IT in Mitigation, Preparedness, Response, and Recovery*. National Research Council. New York: The National Academies Press.

Shklovski, Irina, Moira Burke, Robert Kraut, and Sara Kiesler. 2010. "Technology Adoption and Use in the Aftermath of Hurricane Katrina in New Orleans." *American Behavioral Scientist* 53(8): 1228–46.

Tchamyou, Vanessa. 2017. "The Role of Knowledge Economy in African Business." *Journal of the Knowledge Economy* 8(4): 1189–228.

Temin, Peter. 2017. *The Vanishing Middle Class*. Cambridge, MA: MIT Press.

Therborn, Göran. 1979. *What Does the Ruling Class Do When it Rules?* London: New Left Books.

Thornton, Stuart. 2011. *Real-World Geography: Louisiana Bucket Brigade.* Washington, DC: National Geographic Resource Library.

Tosone, Carol. 2011. "The Legacy of September 11: Shared Trauma, Therapeutic Intimacy, and Professional Posttraumatic Growth." *Traumatology* 17(3): 25–29.

Treisman, Daniel. 2020. "Economic Development and Democracy: Predispositions and Triggers." *Annual Review of Political Science* 23: 241–57.

Ugah, Akobundu. 2007. *Obstacles to Information Access and Use in Developing Countries.* Lincoln, NE: Library Philosophy and Practice.

Universal Declaration of Human Rights (UDHR). 1948. *United Nations.* Retrieved 21 April 2023 from https://www.un.org/en/about-us/universal-declaration-of-human-rights.

Uwihangana, Consoleé, Alexandre Hakizamungu, Charles Rutikanga, and Théogène Bangwanubusa. 2020. *Social Work Practice in Rwanda: Indigenous and Innovative Models of Problem Solving.* Kampala: Fountain Publishers.

Weber, Alan S. 2011. "The Role of Education in Knowledge Economies in Developing Countries." *Procedia: Social and Behavioral Sciences* 15: 2589–94.

Westervelt, Eric. 2014. "Q&A: The Mis-Education of African American Girls." *NPR*, 25 September. Retrieved 21 April 2023 from https://www.npr.org/sections/ed/2014/09/25/351186785/q-a-the-mis-education-of-african-american-girls.

Winthrop, Rebecca. 2010. "Haiti's Earthquake: Prevention and Preparedness Woefully Low." *Brookings*, 14 January. Retrieved 21 April 2023 from https://www.brookings.edu/blog/up-front/2010/01/14/haitis-earthquake-prevention-and-preparedness-woefully-low/.

World Bank Group. 2020. *Open Knowledge Repository: Infratech Policy Toolkit.* Washington, DC: World Bank Institute.

CHAPTER 3

Louisiana

Collaborative Energies and Post-disaster Solutions

JESSICA DANDRIDGE AND YEISHKA MONTALVO

Recognizing the proposition of this text to be advocacy of a conceptual framework of alternative ways in which post-disaster societies of the African diaspora might navigate innovative approaches to achieving sustainable development, this chapter sets out to highlight the inherent value in developing solidarity networks to fight for a world in which dignity, self-determination, and a sustainable future becomes a reality. Emerging post-disaster societies in the US Gulf Coast have found community capital a meaningful way to progress in pursuit of new trajectories for growth and sustainability. The Gulf South is a region where water, oil, gas, coal, and nuclear energy are still key segments of the economy. Louisiana communities, particularly Black, brown, and indigenous, are becoming increasingly vulnerable to the subsequent land loss, rising seas, natural disasters, crumbling infrastructure, and polluting energies weakening an already fragile ecosystem.

The climate crisis has compounded vulnerabilities in Louisiana. In a recent online newsletter on the Louisiana environment, a staff writer posited that an inability to achieve net zero release in greenhouse gas emissions over the next six years would cause Louisiana to suffer damage in excess of $5.5 billion (Schleifstein 2022). In this scenario, the cost of recovery for Louisiana coastal communities would double as a result of climate change. Such observations, regularly shared in environmental newsletters, such as the Catch Basin, that report on environmental issues affecting Louisiana's land, air, and water, bring the deterioration of the Louisiana coastline into stark perspective for residents. Communities have begun to understand what residents would experience in the advent of climate change.

This has prompted a recognition of the urgency for action toward appropriate climate management in Louisiana and become a turning point for solidarity among local communities who have opted to connect their struggles and overcome the systemic inequalities that divide them. Movements and coalitions are united to create a resilient and sustainable future for all; their fight is anchored in the production of knowledge residing in their communities and reposed in the histories, realities, and power of the Gulf region. By advancing grassroots policy and practices that center laborers, farmers, fisher-folk, tribal nations, and frontline communities in a just transition away from extractive economies, sustainable development can be attained. Water management is perhaps the most immediate, but far from the biggest, challenge communities in this region face. Any attempt to confront the challenges for communities in the Gulf region and especially Louisiana, the focus of this chapter, must take into account the aspirations of residents of the region. Those aspirations, at the very heart of this study, lie in the attainment of a better life for vulnerable communities in Louisiana's disaster-prone, diasporic societies.

Seeking Quality of Life

Aristotle spoke a great deal about our attempts as humans to live a "good life." He frequently compared this to an archer's attempt to aim an arrow; the arrow is more likely to hit the target if the archer has a target at which to aim (Wilburn 2015). Similarly, humans are more likely to live a "good life" if they know what makes their lives good. Where and how we aim to hit our target varies depending on the conditions in and circumstances under which we live. In aiming, the archer must look to what suits the occasion, and this involves paying attention to and adjusting for all of the factors that would otherwise make us miss our target (Koehn 2012). What is the direction and strength of crosswinds? Are they constant or intermittent? Are there unexpected moments of turbulence? As these conditions change, the process of hitting the target looks different each time. Are we all aiming at the same target? If so, from what vantage point? Do we all have a fair and equal shot at a "good life"? Is it within our reach (Mendie and Udofia 2020)? These are the questions collaborative efforts in Louisiana must answer.

What do we consider common features of a "good life"? Democracy? Economic development? Human rights? All of the above? If so, then in what order should we initiate action? If these are common features, then why does democracy in advanced industrialized nations work better than in developing nations where growth fluctuates, and political instability is a constant (Gerring 2010)? Perhaps developing nations have been aiming at targets that were never their own in the first place. Maybe they have been following in the footsteps of developed nations, whose conditions are calm and consistent and targets clear. Our attempts to define and determine ways to achieve the "good life" need also to take into account this variety, and it is more complex than archery targets have traditionally been (Wilburn 2015).

Robert Dahl (1989) asserts that a continuing responsiveness of a government to the preferences of its people relies on the government's ability to perceive their preferences as well as act upon them—only then can the "good life" be achieved. In this case, the government is the archer, and the target is their citizen's preferences. Will they take their shot? If they do, will they miss the target entirely? It is possible that they may aim somewhere other than at their primary target; perhaps at a target that is in their own best self-interest and not in the best interest of society at large. How does this scene present itself in post-colonial societies—particularly those of African origin? This chapter will take a look at sustainable development within the context of Louisiana, a society that exists within the borders of the United States of America, a nation that is characterized as the epitome of industrialization and the world's largest national economy. Even then, Louisiana has consistently ranked lowest (#50) in the nation across eight categories: health care, education, economy, infrastructure, opportunity, fiscal stability, crime and corrections, and natural environment (Louisiana Rankings 2022, Jacobs 2023). It is disaster-prone in every way possible. Is it a shock that many have viewed Louisiana in ways similar to their view of a developing nation? If the "good life" as defined by the United States is clear, then why does Louisiana continue to fail and rank last? Why do residents, time and again, continue to miss their target, one they, supposedly, collectively share? How can the Afro-diasporic and other natives of Louisiana acquire the "good life" when their government is not responsive to their needs?

Residents of Louisiana can be better served by striving toward the "good life" as they define it themselves, by being the archers in their story and hitting their own targets. By doing so, they will continue to uncover new ways (or revert to traditional ways) to reform the political systems of oppression that attend development. At the local level, over the past decade, citizens have engaged in innovative, alternative strategies for achieving pressing goals, not just for themselves but also for their communities. They have done so by leaving habits of dependency and crises of legitimacy behind while utilizing their intellectual capacities and reclaiming cultural wisdoms and the traditional energies that have sustained them over generations. Louisiana is culturally different from the rest of the United States with its enduring African, French, Caribbean, Spanish, and indigenous roots and influences (Garrigus and Hall 1994). In knowing their past and leaning into these wisdoms, Louisiana communities have learned how to be resilient, manage catastrophe, and rebuild lives, cultures, and patterns of existence on their own terms.

Louisiana in the Looking Glass

If there were a poster-community for "climate resilience," it would be New Orleans, the cultural capital of Louisiana. Residents of this small metropolis, snuggled along the Mississippi River, bound in every direction by water, have been the definition of resilient in terms of the infrastructure and the painful past and present lives residents have lived. If you were to conduct a search on disaster and resilience, in any database, you would find thousands of news articles, dissertations, books, and podcasts on New Orleans. Many of them carry the same theme of praising residents for their resilient culture. Yet, somehow, one of the most historic cities in the United States has become more closely tied to its trauma than to its recovery. As Judith Rodin, President of The Rockefeller Foundation observed in her keynote address for The Atlantic–Live: "New Orleans is the hot bed, the test bed for all of these [climate resiliency] ideas. It was the springboard for half a billion dollars we've invested into resilience building into cities of all sizes around the world for the last ten years . . . Now it's the template for the hundred resilient cities mode" (Rodin 2015). To become resilient, one has to overcome a hardship or difficulty. It implies that one has learned from a difficulty and grown from it. Residents of Louisiana have gone beyond resilient in many ways and have experienced several moments of collective trauma that have cemented their ability to withstand subpar living standards. Collective trauma, as defined by *Psychology Today* (Turmaud 2022) is the impact of a traumatic experience that can affect a group, community, or nation. Collective trauma is more dangerous than individual trauma because it distresses an entire community, leading to potentially negative consequences that can be built into the fabric of that community. This collective trauma can alter social interactions, political systems, structures, and interactions with others within and beyond the community (Turmaud 2020). In the twenty-first century, the collective trauma felt by residents is now rebranded as resilience, a positive attribute meaning communities can experience more anguish and brutality and bounce back with ferocity. Ironically, the attribute of resilience, collectively or individually, is that it urges the community to assume more than what is expected of another less resilient community. Thus, Louisiana has been the go-to petri dish and lab rat for governments, foundations, engineers, investors, and inventors alike to test good and bad theories because the people's resilience has been established.

For thousands of residents of this city below sea level, this has meant it has been the site of failed experimentation and repeated climate and man-made trauma. The trauma that defines the borders of New Orleans has informed many innovations in music, cuisine, architecture, and even engineering. It has also been the blueprint for failed strategies. Social and economic traumas, as defined by the Trail of Tears, the transatlantic slave trade, classism, neo-liberalism, and oppression have expanded the meaning of resilience to residents of New Orleans. As the unintended perfector of the philosophical concept, residents have taken on the meaning as a description of their self-worth. It is now an identifying marker of the local, the survivor, and the tolerant. To be resilient in New Orleans, you must have lost someone and suffered enough to earn the title. As a result of this proudly worn moniker, residents live in a constant state of recovery. They distrust the disjointed system that continues to underinvest in them and repeatedly fails the community, contributing further to its trauma and civic apathy. In all, New Orleans's resiliency is a project defined by top-down failure of government and neoliberal policy, exposed by the impacts of climate change. More specifically, ravaged by the waters that have created, shaped, and influenced the city, residents have normalized collective trauma and, arguably, find themselves collectively in a constant state of post-traumatic stress disorder.

Louisiana

Disaster in Context

In Louisiana, disaster is presented in numerous formats: super storms, flooding, crumbling infrastructure, government negligence, pandemic, unemployment, corporate greed, and extractive industries, just to name a few. Some maintain that Louisiana is both vulnerable and resilient to disaster, but these disasters are man-made and cannot be detached from the context of colonialism, which created both the economic conditions for disaster and the social conditions that limit Louisiana's capacity to be resilient to it. First and indigenous peoples are chroniclers of their environment, noting and marking memories to events, seasons, and physical changes to their environment (First Peoples Conservation Council 2022). This awareness aids their survival, provoking the concern that resilience and survival may equate to the "good life" in the view of some. The target for vulnerable communities of Louisiana is knowing where and how to build, where to find food, where and how to protect communities, and how to be safe. What happens when the lived environment is changed by disaster, so much so that communities' targets or their sense of the "good life" become unclear? When the landscape changes so rapidly, fishermen can no longer navigate by sight and instead must rely on GPS and radar systems. Leaving the coast for a few months of the year could mean not recognizing certain places when families return. In certain communities outside of New Orleans, residents cannot visit ancestral burial sites because the landscape has undergone devastating change (Forrest 2019).

Western forms of development, with industrialization and the prioritization of extractive industries, have clashed with traditional knowledge and ways of thriving within the lived environment. For example, traditional practices like burning portions of the marshland off the coast of Louisiana each year is now considered dangerous or illegal even though it has been proven to promote healthy regrowth and flush out game for hunting. Other forms of development like oil drilling, agricultural runoff, and flooding have vastly polluted the water while also contaminating the fish. This negatively impacts fishermen, whose livelihood comes from the seafood industry and tribal communities, like the United Houma Nation, whose diets largely consist of fresh fish and homegrown vegetables. In an effort to continue striving for sustainability, tribe members have adjusted their aim and raise beds to garden their crops, cultivate land away from their homes, and purchase fresh produce from others (Forrest 2019).

We can look at extractive industries more broadly and recognize the detrimental effects they have had on climate change and on frontline communities vulnerable to their impacts. These industries are warming the planet and weather events are becoming more and more extreme as the years go by. In the year 2020 alone, Louisiana endured five major storms. It has been eighteen years since Hurricane Katrina (2005), and the devastation is still perceived as third world devastation in a first world country (Bay Area News Group 2016). Victims of extreme weather events like Katrina are overwhelmingly poor, Black, and/or from indigenous communities. These communities have experienced physical and metaphysical damage and the impacts are still seen and felt today. The government and corporations provided very little support in the aftermath and delays in relief and rescue left millions of residents without food, water, or shelter for weeks. This severely undermined the economic stability of these communities in the years since, leaving millions vulnerable to disaster capitalism and private land developers buying up housing stock and displacing thousands of residents from their homes (Adams, Van Hattum, and English 2009).

Hurricane Ida (2021) was a Category 5 storm that occurred exactly sixteen years after Hurricane Katrina, whose impacts were devastating. Louisiana suffered tremendously in its wake. There was no electricity; heat indexes registered a hundred degrees or more; there was no running water or gas; homes were demolished; there was catastrophic flooding, and many residents were trapped in their attics or on their rooftop. The service industry was left without work for weeks without a sense of when they would be able to return to work. Those who had the capacity and resources to evacuate did not know when they would be able to return home or what quality of life they would have, compared to what they once had. Louisiana bayou communities were hit hardest and received very little media coverage or support (Flowers 2021). This was/is not what a government responsive to the needs of its citizens should look like. It begs the question whether a sustainable life is possible or within reach for diasporic communities in Louisiana. Are some communities forced to find ways to thrive by aiming at their own targets with their own arrows? Knowledge production of resident capital can assist such communities to attain their immediate goals by generating homegrown strategies for environmental sustainability.

Beyond super storms and extreme weather events like Katrina and Ida, disasters present themselves in other forms as well. The latter may derive from geographic segregation, generational poverty, lack of accountability for corporate powers and elected leaders perpetuating harm in diasporic communities, historic and current political decisions, and legislative measures that place Black, brown, and indigenous people directly in the zone of disaster, environmental racism, and leaders prioritizing profits over lives. An example of this is the area commonly referred to in the Gulf region as Cancer Alley, an eighty-five-mile stretch of land along the Mississippi River that is known for its nearly 150 oil refineries, plastic plants, and chemical facilities that release cancer-causing emissions into the air, water, and soil (Castellón 2021). Some call these petrochemical plants modern-day plantations because fossil fuel companies have historically exacerbated and exploited segregation. This stretch of land was once the site of plantations with generations of enslaved Africans. Mossville, Louisiana, for example, is a small unincorporated town founded by formerly enslaved people and nearly all of its Black residents have been bought out by the South African petrochemical giant Sasol to build a chemical complex.

Although it has been propagated in some circles that communities of color move into areas where toxic waste sites and landfills are located, there is little to no evidence to support that view. Instead, these sites are often built where poor communities of color have lived for generations. In the 1970s, wealthy white communities became aware of the health risks associated with hazardous facilities and successfully increased their opposition by leveraging their economic and political advantage. Consequently, petrochemical facilities began building their sites near poor communities of color to avoid any delays or expenses associated with building in white affluent neighborhoods. Poor communities of color do not have the same leverage or advantage. GIS mapping has found that polluting industries are in areas with the highest percentages of African Americans, the lowest average household income, and the most residents without a high school diploma (Blodget 2006). Fossil fuel companies exacerbating and exploiting segregation in this way are thus compared to modern-day plantations (Castellón 2021). Black environmentalists in Louisiana are fighting for legislation to protect their communities from racist policies that have worsened their social and environmental burden. The policies that are enacted are ones that fail to address the legacy of racism that contributed to the disasters in the first place. Politicians, too, bear responsibility for failing to enact the same equitable protections and relief for communities of color as for wealthy, white communities. In light of these experiences, it is clear that alternative solutions to sustainable development are urgently required by communities.

Why a Knowledge Economy?

Eighteen years after Hurricane Katrina, and seven years after Ms. Rodin's presentation, a small movement of locals began to redefine "resilient" by creating their own pathways rooted in indigenous and African practices, science, and self-determination. Their focus was on Louisiana's infrastructure. Specifically, the natural infrastructure movement set out to correct and un-engineer centuries of trauma and experimentation, thrust upon residents to navigate without appropriate resources. Before considering opportunities for adapting the infrastructure, however, it is appropriate to explore how the challenge of developing a knowledge economy might lead to a prosperous and effective movement to create resilient communities collaboratively without reliance on governments or foundations for guidance. And while both government and foundations are necessary and have been present throughout the growth of the natural infrastructure movement, for example, residents have pushed external actors to take a lesser role in its expansion. As a result, residents are seeing their quality of life improve through a combination of benefits from green infrastructure and active community participation. The waters surrounding the city have now been co-opted into an opportunity to explore the benefits of knowledge capital and collective action, serving as a physical and metaphysical healing tool that cannot be taught by actors external to the state.

Environmental History of Louisiana

Without reflection on the past, Louisiana's communities are aware that new movements may be doomed to repeat the ills of the past. Here, we will revisit the impact of six climate and man-made events that now determine how a green infrastructure movement may redefine resiliency in Louisiana and harness collective knowledge to recover from trauma. Through a backdrop of repeated trauma, we delve into how residents have conquered these setbacks to become local leaders in green infrastructure, making Louisiana, and especially New Orleans, a projected national leader for climate resilience and natural infrastructure.

Ojibwe of Tears and Trauma

New Orleans has a unique, antiquated infrastructure system that is the most expensive in the United States, making its natural and built environment disjointed, fragmented, and substandard. Because of the city's complex relationship with the land, this city has struggled to balance green space, water, and disaster prevention for centuries. In recent years, the city has taken action to address its painful relationship with the land, water, and its residents. This city has many problems related to green space, but the original problem is rooted in the territory's history. Before the Mississippi River was walled off by levee systems, dams, and industry, it was free to move as it pleased. New Orleans is sometimes referred to as "The Accidental City" (Powell 2013) because it was built by the rich sediment of the Mississippi River over 40 million years ago. Over time, as the river swayed back and forth across the coastal plain, Louisiana spanned approximately five thousand years and was the last land formation before reaching the Gulf of Mexico, known as the Mississippi Embayment. Connecting waterways from all over the country, the latter moved sediment, silt, and sand to form Southeast Louisiana (Severin et al. n.d.). To give context to the abilities of the river to shape and create land, the waters of the Mississippi River were so strong and rich that they covered a small barrier island of sand known as Pine Island, which is now the base of New Orleans (Campanella 2016). The subtropical environment of the Louisiana Delta also meant that the ecosystem was diverse and plentiful. This made it a prime spot for over twenty indigenous tribes who traded goods, supplies, food, and medicines up and down the Mississippi River Valley, known to the indigenous tribes as Ojibwe or "Great River." New Orleans became known as *Blvbancha* or "land of many tongues" by the Chitimacha Tribe because of how diverse, vibrant, and plentiful the lands were (Burton, Smith, and Appleford n.d.; "Yellow Fever" 2022; Levenson 2018). The indigenous tribes of Louisiana understood that the lushness of the Delta came with a set of unique ecological expectations. Tribes flourished as they worked with the Mississippi's strong currents, understanding when it would flood and when it would plateau. They knew when certain species of birds, mammals, fish, and insects would be copious or sparse, based on the season. The largest indigenous mounds and historical sites can be found from the headwaters in Minnesota to Louisiana. Because of how plentiful resources were, they thrived undisturbed by outsiders until the 1500s (Burton et al. n.d.).

The Mississippi River Valley was first discovered by Europeans in 1541, pinning the beginning of the history of racial and climate trauma in Louisiana (Johnson, Fry & Co. 1858). French explorer René-Robert Cavelier, sieur (lord) de La Salle claimed the Delta for France, giving Louisiana its name after Louis XIV and his wife Anna. La Salle and other French businessmen saw the strategic importance of controlling the Mississippi River. The ability to control the waters of the Mississippi meant controlling commerce between the Americans to the East and the Spanish to the West. Over the next 300 years, indigenous tribes battling for their ancestral lands were enslaved or died of disease brought over by the Europeans. The Indian Removal Act of 1830 marked the final stage of the nineteenth-century vision for the expansion of the transatlantic slave trade, which had already begun in 1619 (Hannah-Jones 2019). When Louisiana was claimed by the French in 1682, they had brought slaves with them from Africa and other colonies such as Haiti. From 1619 to 1830, the French, as did all other European powers of the time, embedded slavery into their economy, culture, and social systems. When Thomas Jefferson purchased Louisiana on 30 April 1803, the United States not only acquired Louisiana but also the entire Mississippi River Valley and the central plains, thus allowing for further expansion of the United States, human conquest, and conquest of the river's economic potential (De Cesar and Page, 2003). Alexis de Tocqueville, French aristocrat, historian, and political philosopher witnessed the trauma firsthand and lamented the fate of the Indians, who had their families with them and brought in their train the wounded and sick, with children newly born, and old men upon the verge of death. The Indians, without tents or wagons, and very few provisions, embarked to pass the mighty river, in silence, knowing their calamity to be irremediable (Bill of Rights Institute n.d.).

This moment of national and localized trauma was the first of many. For the Delta region, this meant that indigenous knowledge of the land, waterways, and its lush ecosystem would, in many cases, be lost for centuries if not forever. Some indigenous tribes were able to elude white settlers and others signed deals that allowed them to keep their lands if they cooperated. This cooperation meant assimilation, isolation, and indefinite oppression for all tribes of the once thriving Mississippi River Valley. Today, the Chitimacha Tribe is the only tribe still living on a portion of their ancestral lands.

The Transatlantic Slave Trade

While indigenous tribes were being removed or killed, the transatlantic slave trade was expanding, making the need for indigenous land more valuable, driven by the thriving plantation economy. Thanks to the Mississippi River's fertile land and subtropical climate perfect for crop growing and its access to major waterways for trading and selling, the Mississippi, and surrounding lakes, bayous, and swamps made New Orleans the perfect place for white male business owners to prosper. Innovations such as the cotton gin, creole sugarcane, and the steamboat made New Orleans and Louisiana an economic powerhouse throughout the time of slavery. Several years after New Orleans was founded in 1718, it became the busiest slave market in the south. More than 135,000 enslaved Africans were sold in the heart of the French Quarter. This provided the labor for hundreds of plantations along the banks of the Mississippi River. Products cultivated by the enslaved included sugar, indigo, and cotton. By 1860, almost half of the Louisiana population comprised the enslaved. Among their many tasks, enslaved Africans had to cultivate crops, tend to livestock, and farm food for their own families as well as the plantation owners' families. Enslaved families also had to build plantation homes, stables, and all other forms of infrastructure common in the United States. What was unique in Louisiana was that the enslaved had to build levees and other forms of drainage, which was considered the hardest and most brutal work of the time (Louisiana State Museum n.d.).

Plantations were strategically built near waterways to move and sell product faster. Plantations, farms, and other industries in the Delta region alongside smaller rivers and lakes were prone to flooding. This meant that the enslaved were charged with building complex levees and drainage systems to prevent or reduce damage to their homes and products. It was no small feat to build drainage. "Levees required extensive upkeep and were under constant surveillance for seepage and weakness. Planters tasked their slaves with hauling clay for structural reinforcement, tending the grassy slopes, clearing clogged sluices and ditches, and cutting trees whose root systems would damage the levees' structural integrity" (Davis 2018). For many of the enslaved, working the levees was worse than being in a chain gang and meant endless days and nights of hard labor. To make matters worse, these levees were often not led by engineers or architects.

Plantation owners or politicians gave orders to the enslaved. Almost every spring, when the river was high, or after a storm, the levees would fail or sometimes be sabotaged by competing plantations. Failed levees meant that crops protected by the levee systems, also failed. Thus, the enslaved were not only further oppressed by their owners and the government but they were subject to the constant whims of nature. It was in the slaves' own interest to ensure the structural soundness of the levees and drainage lest they face more backbreaking labor in the fall and winter (Rodin 2015). Work songs to this effect have been recorded.

While the Mississippi River could serve as a means of escape to the underground railroad, or through the swamps into hard-to-find indigenous tribes known as Maroons, the collective memories of the enslaved in Louisiana was shaped by the river's entanglement with white colonizers' desire to conquer the river, the land, and the people simultaneously. Because indigenous tribes and the enslaved could not read or write stories of collective trauma, their narratives were passed down for generations. Collective trauma was then written and ingrained into local culture, sealing separation between Black communities and nature for over a century.

Sinking Ambitions of Conquest

After the Emancipation Proclamation in 1863 and the fall of the Confederacy in 1865, Louisiana saw rapid change that brought on what was believed to be massive infrastructural and economic advancements that would later cause more collective trauma and ended residents' true ability to have autonomy over their land and their communities. Louisiana was exploited for its rich sediment and New Orleans was where the deals for various forms of exploitation were signed. In quick succession, these events took place throughout the twentieth century. Four moments defined Louisiana's relationship with the natural ecology and hydrology of the Delta that remained mostly untouched until the twentieth century. The quick succession of these events provided short term benefits and, as we will discuss, long-term and irreversible negative impacts on the environment, political will, and social upward mobility.

Prior to the emancipation of the enslaved and after, the city and state struggled with drainage. Even with thousands of slaves working around the clock, the problems of proper drainage, sewerage, and sanitation were difficult to manage, especially considering the city's natural topography. Most towns and economic centers in Louisiana were built by the Mississippi River on ancient ridges. Nevertheless, Louisiana still saw consistent flooding because the land was built by the rich sediment from the river and nearby lakes and bayous. At this time, technology and engineering could not hold back the power of the Mississippi (Davis 2018). Architectural design at the time was pulled directly from French and Spanish architecture and did not adapt to seasonal flooding and subtropical rain and storms. Additionally, drinking water, which came from the Mississippi River, was unsanitary and caused a variety of illnesses because residents could not figure out how to properly sanitize the muddy Mississippi waters. According to Emily Perkins and John Magill of the Historic New Orleans Collection, over 4,000 people died from yellow fever in 1878 caused by the lack of pavement and drainage. In 1880, over 80 percent of the city remained unpaved. A virus, caused by mosquitoes from Africa and Latin America, took effect within six days and caused serious illness and death. It was unclear to residents in the nineteenth century what was causing the massive death in the wet summer months but everyone knew something must be done. The same year that yellow fever killed over 4,000 people in New Orleans, residents also learned that over 20,000 died along the Mississippi River Valley all the way to Memphis from the yellow fever that had started in New Orleans (Perkins and Magill 2020).

As a response to these devasting events, travel and trade to New Orleans was banned and cities along the Mississippi closed their ports to ships coming out of New Orleans. After several failed millage attempts to pay for a new drainage system, residents finally agreed to a new millage in 1899 leading to the development of the Sewerage and Water Board (Perkins and Magill 2020). By 1905, the city finally built its first water sanitation plant and by 1913, the wood screw pump was invented, which catalyzed a new movement and forever changed the landscape of New Orleans and all of Southeast Louisiana. Designed by a local engineer in 1913, the wood screw pump became the leading method to rid the city of standing water and consistent flooding forever, reducing the risk of yellow fever and allowing for the expansion of the Orleans Parish and other coastal parishes. To do this, healthy marshland was drained to make way for development of post–World War II construction. This invention, resulting from location of the knowledge capital in local communities, was the key to the city's cultural, economic, and structural expansion so it could become a major metropolis of the South. As a result, trees, grass, parks, and other characteristics of the natural environment were quickly removed and paved for development (Lux 2018). Fear of another yellow fever outbreak instilled a deep resistance to standing water, and greenspace made way for an unprecedented amount of pavement. Additionally, this work contributed to rapid subsidence in New Orleans and surrounding parishes. Suddenly, the spongy Delta was forcibly dried out, pushing the land downward and making it more prone to flooding. As is often noted, no good deed goes unpunished. In finding a solution to the spread of disease and other homegrown fixes for the infrastructure, new vulnerabilities were exposed.

Oil Wealth

As New Orleans was grappling with the struggling drainage systems plaguing residents up and down river, the oil boom was revving up. The first oil well discovered in Louisiana was found in Jennings, Louisiana in 1901. The small well produced over 7,000 barrels a day and exploration across North and Central Louisiana commenced (Wells and Wells 2005). While the story of oil in Louisiana is ongoing, the most important detail related to it was that the industry advanced technology, with the marsh buggy and drill barges that allowed for ongoing oil exploration in water and coastal areas. By the 1930s, oil fields in Southeast Louisiana dwarfed fields in northern Louisiana, which were much smaller and, in many cases, already emptied. By the 1970s, most oil drilling was located offshore in the deepest parts of the Gulf of Mexico. Beyond oil drilling and processing being toxic to the environment, the greatest threat to Louisiana was the ongoing destruction of coastal lands. To reach oil in the deep waters of the Gulf of Mexico, developers built canals through freshwater marshes and swamps, leading to irreversible saltwater intrusion (Phillips 2020). Over 10,000 miles of canal were cut into the once healthy marshland, which could stretch halfway around the world. To make matters worse, over 13,000 miles of debris from the development of these canals sat where they were left, causing further damage and killing the wetlands with either too much or too little water.

The big oil boom was great for the economy until the 1980s, and at the height of the boom, oil companies were pumping 360 million barrels a day. Roughly an eighth of what is extracted from Saudi Arabia was being harvested from the small and impoverished state of Louisiana. The toxic brine, a water byproduct with ten times the salinity of the ocean, was dumped into marsh lands and waterways killing any leftover land and polluting rivers and bayous from which residents thrived in their hunt for food (Meehan et al. 2017). What was good for the few oil investors was bad for everyone else in Louisiana and by the time brine was regulated in 1985, the damage was already done. The damage between saltwater intrusion, subsidence and toxic oil and gas brine, is ongoing and Louisiana loses an estimated football field a day of coastal land due to saltwater intrusion and sea level rise. Louisiana is estimated to lose roughly $15 billion per year to ecological and economic development that could otherwise be profitable for a variety of other industries (Burdreau 2018; Mahoney 2021).

The Great Flood

Another watershed moment for Louisiana occurred in the 1920s. It altered the fabric of the nation. One writer for the *New York Times* argued that "once more, war was on between the mighty old dragon that is the Mississippi River and his ancient enemy, man" (quoted in Parrish 2017). Heavy rain and snow from the Ohio River Valley rushed downriver toward the Mississippi River bursting levees, starting in Illinois to Southeast Louisiana, from March to May of 1927. During that time, 27,000 square miles of land were inundated with water breaking all the man-made levee systems and destroying homes and communities. Over 170 counties in seven states were severely impacted. Within that time, an estimated 637,000 to 931,159 people became homeless, with an estimated 555,000 of these individuals being racial and ethnic minorities. In Louisiana alone, over 10,000 square miles across twenty parishes were submerged (Bradshaw 2011). Additionally, New Orleans saw record rainfall of up to fourteen inches in April (1927) during the Great Flood. While the levees did not fail, they were expected to do so within days. The only solution for protecting the city and Baton Rouge from mass destruction was to blast the levees downriver. That would mean flooding an entire area, on purpose, to save the major economic and political areas. Even though communities downriver protested, the governor and the federal government proceeded. Over the next ten days the National Guard blew the levees at Caenarvon Township using 39 tons of explosives on the levees and releasing over 250,000 cubic feet of water per second into neighborhoods and farmland. Over 10,000 people overnight became refugees and were moved temporarily to New Orleans while they waited for the flood waters to recede. Each family was supposed to be compensated, but they grossly underestimated the financial impact. Some families received a total of $274, while most, especially Black families, received nothing to this day. The same could be said west of Louisiana where more rural farmlands were flooded. Nearly 60,000 families were displaced but many saw no compensation or support to help rebuild. To add to the mass tragedy, the refugee camps also perpetuated extreme racial inequality. Supplies and means of evacuation after flooding for both those in Caenarvon and the Atchafalaya River Basin were given strictly to white citizens, with African Americans receiving only those supplies that were left. African Americans also did not receive supplies without providing the name of their white employer or a voucher from a white person. White supremacists saw this as an opportunity to further exploit Black labor. African Americans were frequently forced to work against their will and were not permitted to leave the camps to care for and tend to their land and families. The Red Cross, along with Secretary of

Commerce Herbert Hoover, deliberately concealed the abuses that Black refugees suffered in Red Cross camps in order to obtain donations for the rebuilding effort. Thus continued the oppressive subjugation of diasporic peoples in the industrialized world (Bradshaw 2011; Parrish 2017).

Hurricanes

The sixth historic *marker* to impact Louisiana is its hurricane experience. This climatic history begins with the biggest legacy of the flood of 1927: the creation of the Army Corp of Engineers. In 1928, the Flood Control Act was passed by President Coolidge, creating the first ever flood control system managing the entire length of the Mississippi River and connecting rivers and tributaries. The new federal agency received $325 million to design, engineer, and build levees, spillways, dams, and diversion channels. For nearly seventy-eight years, the levee systems prevented massive flooding events like the flood of 1927 and previous flooding events.

On 9 September 1965, Hurricane Betsy made landfall as a Category 3 storm and challenged man-made engineering. The Mississippi River Gulf Outlet (MRGO), a 76-mile-long navigation channel was built by the Port of New Orleans and the Army Corp of Engineers. Built to be a "shortcut" to the Gulf of Mexico for ships bringing goods in and out of the country, it soon became clear that it was instead a shortcut for storm surge flooding the Lower 9th Ward Neighborhood of New Orleans and St. Bernard Parish. Flooding over 165,000 homes of mainly Black and low-income communities and destroying coastal communities and farmland, Hurricane Betsy became the first flood to top $1 billion in damages (Ouchley 2015). As a response, the Army Corps designed the Hurricane Protection Program to protect the city of New Orleans against Category 5 storms. Unfortunately, on 29 August 2005, the Army Corps levees failed yet again, devasting the entire Southeast Louisiana Region, and parts of Mississippi and Alabama coastal communities.

On 29 August 2005, Hurricane Katrina proved to be more than just a case of failed levees for New Orleans and parts of Southeast Louisiana. It was a collision of failed government on all levels and failed engineering dating back to the nineteenth century. Nearly two centuries had passed and, once more, man-made engineering failed residents. Diasporic and indigenous communities endured the brunt of the pain, stress, and trauma. After Hurricane Betsy, the Army Corp had opted not to close and fill in the MRGO, leaving residents of New Orleans East and the Lower 9th and St. Bernard Parish vulnerable to massive flooding. After the Hurricane Protection Levee System was built, the authorities failed to maintain them year after year, as the slaves and poor were forced to do yearly due to natural erosion. After a hundred years of practice, New Orleans' leadership still failed to protect the poor, elderly, and disabled residents in at-risk areas. Further, the local government failed to provide residents with food, water, and supplies days after the storm passed. Once more, the state and federal government neglected to provide federal assistance to residents who were again subject to the whims of poor government planning. Residents of Southeast Louisiana did not receive proper assistance for the loss of their lands despite their payment into flood insurance programs and the oversight of FEMA and other federal agencies charged with protecting residents. When the recovery process began, many realized that Hurricane Katrina further exacerbated issues left unnoticed or untouched over past decades.

The levee systems, managed by the Levee Board and maintained by the Army Corps of Engineers, failed in fifty different locations, flooding over 80 percent of the city. Eighteen years and $25 billion later, significant initiatives have been taken, through federal investment, to repair gray infrastructure as part of the recovery process. This was a clear example of adding funds to benefit a failing system, rather than caring for the residents who were most impacted by its failure. This further exacerbated the inequities. The yellow fever outbreak and the installation of the new drainage system that had already left New Orleans without ample greenspace brought further devastation with Hurricane Katrina, which stripped the city of its trees and green space. The failure of the pumps and the failure of the levees led to mass flooding of a large portion of the city. Toxic standing water remained for over six weeks and led to the mass death of native plants, trees, and grass. When the water was drained, the city lost 60 percent of its already sparse tree canopy. The recovery perpetuated these problems because many homeowners did not receive funding to replace trees or plants.

The hurricane further underlined greenspace inequities because wealthier white neighborhoods with disposable income for landscaping improvement saw far less flooding on higher land. Finally, Black homeowners who could not receive recovery dollars or those who chose not to return left their properties untouched, a sight still evidenced by homes with water lines or search and rescue symbols. In 2019, the city estimated that there were thirty thousand blighted homes still peppering the city, with most concentrated in Black neighborhoods. This blight is more than unsightly; it is a breeding ground for invasive species and encourages divestment and crime (Monteverde 2021). Today, these issues have created the perfect storm as climate change has put New Orleans at increased risk for greater pluvial flooding and tropical storms. In 2021, Louisiana saw the highest level of rainfall in recorded history along with several of the strongest storms ever recorded (Baurick and Parker, 2021; Broach 2022). The pumping system, built over a hundred years ago, was not created for the frequency and intensity of the rain, making the system a financial risk and burden rather than an asset. The lack of trees and green space compounded with subsidence exacerbates the impacts of these storms. Eighteen years since Katrina, New Orleans is still grappling with the loss of green space from over 300 years ago.

In post-Katrina recovery, which many would argue Southeast Louisiana is still engaging in, residents are often perceived to be resilient. Harking back to Judith Rodin's presentation on behalf of the Rockefeller Foundation, she noted that "storms may be inevitable, but crisis is not. Crisis does not have to become disaster and by building resilience, New Orleans and other cities can prepare for the next disruption while building a stronger society and economy at the same time." She further pointed out that "when a misfortune hits, as it did in New Orleans, cities have the opportunity to grow, to transform, and this is the essence of resilience building" (Rodin 2015). It is widely recognized that resilient communities are as aware of their assets as their vulnerabilities. Through coalition building, communities can build capacity and self-regulate to become adaptive and flexible. By locating their assets and integrating their intellectual capital, they can engage in knowledge-sharing to avoid redundancy and advocate for environmental sustainability.

Knowledge Approaches to Sustainable Development

As Louisiana engages strategic recovery planning, recognizing disaster, attributed to Mother Nature, man-made errors, destructive behavioral patterns, myopic elected officials, underfunded government agencies, and inadequately trained engineers, budding architects of recovery planning realize that the gray infrastructure, pivotal to the city's success in the twentieth century, is vulnerable to the vicissitudes of the twenty-first century and the intensification of climate change. As local communities and leadership begin to seek recovery from poor systems, they realize that there were several core problems with how Europeans and Americans tackled the natural infrastructure of Louisiana and, specifically, New Orleans.

First, engineers have tried, over decades, to prevent the Delta from doing what it was meant to do. The Louisiana Delta only exists today because the Mississippi River flooded yearly for thousands of years, pouring rich sediment, sand, and silt over each other, developing a spongy, soft land mass that requires frequent regeneration from the river to replenish itself, lest the land sink. After three hundred years since the founding of New

Orleans, it should be finally realized that humans cannot control how the natural ecology works. Especially considering that gray infrastructure and the built environment, meant to hold the river and other bodies of water back, were only temporary solutions.

Second, Louisiana realized that there must be a better, more practical way to define what "resiliency" means. The people of Louisiana have been called resilient for their ability to bounce back and recover quickly from tragedy, misfortune, or change. But, considering the lack of infrastructure or systems to prevent the repetition of these tragedies, one could question whether Louisiana's residents are indeed resilient or just accommodating. One of the best commentaries on resilience is found in an article entitled "A Framework to Quantitatively Assess and Enhance the Seismic Resilience of Communities" (Bruneau et al. 2003) where resilience is perceived as "the ability of social units (e.g., organizations, communities) to mitigate hazards, contain the effects of disasters when they occur, and carry out recovery activities in ways that minimize social disruption and mitigate the effects of future earthquakes." To enhance seismic resilience, therefore, is to minimize loss of life, injuries, and other economic losses for the benefit of preserving quality of life. The relevance of this observation to disaster in Louisiana is clear. Responding to hurricanes should share similar objectives of minimizing loss of life, injuries, economic losses, and a reduction to the quality of human life. This recognition, over time, has led to new movements to re-develop Louisiana and engage collaborative energies within new, community-based coalitions for post-disaster recovery. The progress made by these collaborations constitute a significant step toward a true embrace of resilience and efforts aimed at redesigning green infrastructure, redefining urban water management, and envisioning resiliency in terms of achieving a better quality of life.

Third, the last core issue rarely addressed is collective trauma due to ongoing racial and environmental injustice. While the ability for residents of Louisiana to endure numerous tragedies can still be seen as an asset, it can lead to political apathy and distrust. On a political and economic level, it has prompted the federal government to invest less in a community knowing that they can recover without the required tools and resources. Furthermore, it has held the state back, leading to generational poverty, reduced economic opportunities, massive wealth gaps, and deep poverty for Black and indigenous communities still recovering from events prior to the twentieth century. In 2019, Louisiana ranked fiftieth in the union overall, forty-eighth in education, forty-eighth in opportunity, forty-second in fiscal stability, and fiftieth in natural environment ("Overview of Louisiana" 2021). These rankings are rooted in racial inequality and weighted by the history of environmental injustice and climate degradation. All the above outcomes are deeply ingrained in colonization, and the belief that the land and people, as well as their outputs must be controlled and monopolized for economic opportunity, even if that risks the failure of social and ecological systems. The very idea that the largest river in the United States must be controlled by humans is absurd and has led to natural disasters, furthering man-made error and state sanctioned violence.

Post-Disaster Recovery

Infrastructure

The natural infrastructure movement was launched in 2014 by a small group of kindred spirits. A framework for development was laid out in four dimensions:

1. Robustness: the strength of the system to withstand a given level of stress.
2. Redundancy: the extent to which a system can maintain its functional requirements in the event of disruption.
3. Resourcefulness: the capacity to identify problems, challenges, and mobilize resources as needed to prevent degradation of the system.
4. Rapidity: the ability to mobilize resources in a timely manner to contain losses and reduce further disruption.

The above dimensions are seen as hallmarks of resiliency. Since many would agree that Louisiana is not nearly as resilient as it is anecdotally described, redesigning for resilience begins with an understanding of what is needed to claim successful management of Louisiana's most central issues. A framework for developing the local infrastructure underlies what it would take for Louisiana to ground its recovery and claim its best life. It may be true that while residents display high levels of perseverance, endurance, and tenacity, being unable to overcome high levels of injustice, oppression, and inequity resulting from systematic challenges of inept government on all levels has left Louisiana residents with a tolerance for the status quo. But resiliency requires more than tolerance and the ability to adapt. So, what would it take to achieve true recovery, claim every dimension of resiliency, and transform the infrastructure to meet the challenges of a changing global climate? Identifying new ways of knowing and producing knowledge for a better quality of life is a good starting point for Louisiana communities and collaborative engagement of the knowledge resources available to them.

After centuries of error, a group of professionals, realized that moving forward with the tried methodology would be continuing a broken system, and constitute a waste of taxpayer dollars. Founders of The Water Collaborative of Greater New Orleans understood that New Orleans required a new methodology to combat the bowl effect and the mounting inequities from pre- and post-Katrina. Motivated by meetings with the Dutch and ongoing research, they realized the necessity to un-engineer the engineered and live with water, a core component of this quasi-grassroots movement. In 2006, a year after Hurricane Katrina, the Dutch Embassy invited a delegation of Louisiana residents to learn about how the Dutch live with water. Two sets of meetings took place in 2008 ("Dutch Dialogues" 2014). The root of the water management movement emanated from the Dutch dialogues, which brought multiple disciplines together to solve resiliency and risk mitigation issues. "As opposed to working in silos, the dialogues allow people to work across disciplines to work on problems as the climate changes. Those challenges could be about drought, flood, or water supply issues" ("Dutch Dialogues" 2014).

David Waggoner, regarded by many as the godfather of the "living with water" movement in Louisiana was part of the 2006 Louisiana delegation and developed the Dutch dialogues. He knew that Louisiana's gray infrastructure systems were failing but our values and processes around these systems were also an essential problem, something that had not been discussed beyond a theoretical level. Living with water, is more than an engineering technique; it is a way of life and a cultural shift. Concepts that go beyond problem solving but are deeply rooted in design thinking, systems integration, and interdisciplinary approaches are closely tied to the success of the Dutch. Unlike the Dutch, Louisiana's water systems are rooted in racism and classism, and caused generational traumas brought on by various forms of violence, making the challenges much more difficult to address.

In 2010, David Waggoner received federal funding from the state office of Community Development to turn ideas into designs and plans. Waggoner & Ball and a team of designers developed a comprehensive plan to transform New Orleans, known as the Greater New Orleans Urban Water Plan, which was published in 2014. While they tackled various issues and presented it beautifully and clearly, it could also be considered a book of philosophy because government officials did not buy into the concepts put forth initially (Waggonner & Ball n.d. [2014]). Waggoner and founding members of The Water Collaborative understood that it would take more than a beautiful book, it required action. Thus, The Water Collaborative (TWC) was built by participants in the Dutch dialogues and the Urban Water Plan. This process, from the Delegation to the Urban Water Plan, took fourteen years to get from concept to action (Waggonner & Ball n.d. [2014]). In 2015, a strategic plan funded by the Kresge Foundation, for The Water Collaborative was finalized. At the core was still the physical as well as cultural shift of "living with water."

For New Orleans, living and thriving with water means making the most of the city's surface-built environment pervious to reduce natural and man-made flooding risk, undo historic racist practices, and create a durable city that could sustain stronger tropical storms, pluvial flooding, and other climate risks such as heatwaves and drought. This vision also includes a city that thrives by addressing climate, racial, and class trauma through the development of green spaces. To achieve this, the core concept was to make New Orleans a sponge city again and return it to its Delta state. Sponge city is exactly as it sounds. Since the Louisiana Delta is made of a mix of water, sand, silt, and other sediment, it makes sense to return the land to its spongy nature. To achieve this, natural infrastructure, or specifically green infrastructure would be implemented to scale. From residential to public and private land, water would need to be sunk, slowed, or stored to better live with it and create true resiliency.

The Infrastructure Movement

Before we go into the details of this movement, it is important to define green infrastructure as well as natural infrastructure. There are several definitions of natural infrastructure but there are two that truly stand out. The first definition by the International Union for Conservation of Nature, the World Bank Group, and the World Resource Institute, represents nature-based solutions as: "Actions to protect, sustainably manage, and restore natural or modified ecosystems that address societal challenges effectively and adaptively, simultaneously providing human well-being and biodiversity benefits" (Luedke 2019).

Another definition that speaks to the core ideas of the natural infrastructure movement in Louisiana is the US Army Corp of Engineer's definition. Ironically after almost a hundred years of gray infrastructure development across the United States, the Corps has come to terms with the benefits of natural infrastructure and the shortcomings of gray infrastructure. They define natural infrastructure as engineering with nature or "the intentional alignment of natural and engineering processes to efficiently and sustainably deliver economic, environmental, and social benefits through collaboration" (Bridges et al. 2021).

Basically, natural infrastructure in its truest form is just building back nature that has been eroded through man-made development, natural erosion or climate change examples of natural infrastructure include living shorelines, sand dunes, coral reefs, coastal wetlands, nature reserves, and even urban forests are part of the natural infrastructure movement (Luedke 2019).

Green infrastructure is very similar and falls under the category of natural infrastructure although the difference is that green infrastructure incorporates gray infrastructure systems as part of its design since urban or suburban environments should be connected to other systems such as sewerage and drainage lines and a transportation system. Further, the American Society of Landscape Architects define green infrastructure as: "Humans harnessing nature for use as infrastructural systems. This can be done at a landscape or site-specific scale and includes both natural systems restoration or protection and build projects that emulate nature" (Luedke 2019).

Types of green infrastructure include raingardens, bioswales, retention ponds, green roofs and walls, tree cells, and even rain barrels or cisterns. These ideas remained insular and within design groups in the city. The natural infrastructure movement picked up steam after two major events in Southeast Louisiana. The first event occurred in 2016 and in multiple parishes: Baton Rouge, Ascension, Livingston, and Tangipahoa parishes experienced heavy rainfall and flooding. What was predicted to be a 6–10-inch rain event turned into a 26-inch rain event and flooded the area with three days of nonstop rain. As a result, thirteen people died, 100,000 homes were destroyed with an estimated 8.7 billion in damages (Samuels 2017). The second event took place in 2017 in New Orleans on another unsuspecting day. The forecast for that day expected rain but did not predict the heavy downpour that would inundate the city with 3–5 inches of rainfall in less than an hour and nearly 9.5 inches in five hours, overwhelming the hundred-year-old drainage system (Gabour 2017). On a good day, the pumping stations, which can pump roughly an inch in the first hour and a half and another inch every hour thereafter, could not keep up with the downpour. What was later disclosed was that several of the pumps were not in service, leaving the city helpless to flooding rain. For nearly half a day, the city was sitting in water, ranging from ankle deep to knee deep. Residents even pulled out their canoes and boats to get around town (Gabour 2017).

After that event, elected officials were frantic to find solutions. Options were to upgrade and put more resources into the aging infrastructure systems. Many realized that the gray infrastructure systems were part of the problem. Thus, the need for green infrastructure to manage water became a priority for both elected and business leaders. The timing was perfect for The Water Collaborative, which became incorporated that same year. Other movements also sprang into action. In 2015, the Gentilly Resiliency District had won $141.2 million through the US Department of Housing and Urban Development (HUD) for retrofitting Gentilly, a low-lying, residential neighborhood of New Orleans into a fully integrated water resilient community. This would be the first shot at city government putting major dollars into the concept of living with water. Furthermore, many of the companies and nonprofits that participated in the development of the Urban Water Plan had already begun integrating green infrastructure on private projects ("Fiscal Year (FY) 2014 Funds National Disaster Resilience Competition" 2014). The cultural shift discussed in the Dutch dialogues had yet to become integrated in social, cultural, and political systems. It did not take long, though, for the shift to begin. The movement of green infrastructure in collaboration with businesses and nonprofits worked swiftly to retrofit what was once marsh and swampland into a sponge city.

Achievements of the Infrastructure Movement

Some may wonder how a landscape architectural method for managing water would be as successful as Louisiana's became in less than a year. Since the beginning of the settlement of the land in the Mississippi Delta region by Europeans, there has been a focus on redeveloping the natural ecology of Louisiana into something that it was not, causing continual failure of gray infrastructure systems and collective trauma within Black, indigenous, and low-income communities across the state. The benefits of natural and green infrastructure are immense and include the reduction, or mitigation of urban, suburban, and rural flooding, reduction in heat island effects, and a form of reduction carbon strategy through natural carbon sequestration. Green infrastructure also improves overall quality of life through beautification of communities, increased property values, and space for community gatherings and exercise. Finally, improving a community's overall ability to withstand the impacts of climate change such as sea level rise and extreme weather events, marks the success of the collaboration. This plethora of benefits create true resiliency allowing communities to bounce back faster after major events (US EPA 2022). In terms of resiliency, the four dimensions were easily met and helped move a community, such as the Greater New Orleans area, from tenacity and perseverance to resourcefulness and resilience. More so, the shift to green infrastructure allows residents to see the direct benefit of climate change adaptation strategies that are often not available or accessible to the average resident.

Most residents of Southeast Louisiana are dealing with both the impacts of climate change as well as the ongoing impacts of collective trauma. In terms of collective trauma, many residents do not trust the infrastructure systems because many are feeling the impacts of Hurricane Katrina, Hurricane Betsy, and even the flood of 1927, which still impact thousands across the state. Often gray infrastructure is a top-down approach authorized and managed by government agencies whom local communities will never know. Nor will residents participate in the design developmental stages, creating systems that perpetuate displacement, causing additional stress to residents when these agencies should be providing relief. Gray infrastructure often does not include community voice, thus taking residents' autonomy and self-determination away from them and leaving them helpless when designs are implemented. Natural infrastructure or green infrastructure creates a new pathway for developing systems that are bottom-up and incorporate community participation, perspective, and voice throughout the entire process. For Louisiana, this is a game changer and not only allows residents to create more opportunities for their own self-determination on their own land or property, but they also allow residents to see firsthand the positive impacts of climate mitigation and adaptation strategies. Natural infrastructure strategies, such as rain gardens, French drains, or even the simple rain barrel, provide residents with real world examples of what the future of Louisiana could look like.

From the perspective of economic impact, it creates new jobs and a robust workforce that employs even entry level workers with adequate disposable income. Yet, it does not end with construction jobs, it is also an intersectional discipline where one can come from many different sectors and have a voice on how green infrastructure is implemented, designed, taught, and incorporated in our day-to-day policies. From urban planners, policy writers, educators, horticulturalists, urban forestry, urban farming, to wildlife management and fisheries, all have a place in natural infrastructure (Greater New Orleans Foundation 2021). Members of The Water Collaborative understood that to see large scale change in how the state deals with the impacts of climate change, collective trauma, and environmental injustices, spaces must be found for intersectional design thinking to create sustainable solutions. Green infrastructure offers a unique opportunity to bring people together from various backgrounds, disciplines, ethnicities, gender identities, and classes around a shared goal. At the core of the great infrastructure or natural infrastructure movement is the idea that we are not only changing the physical infrastructure but the social infrastructure as well. The green infrastructure movement is a social change movement that requires the cultural shift of residents as well as elected leadership. To change the way our system works, we must change how we prioritize certain aspects of our ecology and change how we respond to the natural infrastructure and environment on a day-to-day basis. For this change to occur, collaboration must not expel or dismiss any individual's vision or values. It also requires flexible and impartial leadership because often perspectives on what resiliency or sustainability means to us personally is rooted in our own personal trauma, behaviors, thought patterns, and socialization. One of the motivational questions we may each ask ourselves is:

> If I work in or adjacent to the non-profit and movement ecosystem, how am I holding on to purpose, especially given the pulls and contradictions of purity (having to prove that we belong in movement spaces), productivity (maintaining constant performance and output at the cost of overwork, replication, and low impact), personality (pursuing relevance and celebrity activism), and perfectionism (expecting to get it right all the time and being afraid to own up to mistakes)? (Lyer 2020)

For those working in the nonprofit or movement ecosystem or building structures or systems toward environmental justice, we are often bogged down with the many contradictions of purity, productivity personality and perfectionism, often rooted in colonialist thought practices. We often do not realize that when we start moving toward larger system change, we are also working from a colonized framework that can often be in direct opposition to the work we are trying to instill in our own communities. Progress can only occur when one includes the many complex voices even when some of those voices are out of sync, especially in relation to economic development, jobs, gentrification, or public health. For this reason, collaborating in diasporic spaces around climate adaptation and mitigation can motivate and inspire those in less developed regions who have encountered similar silencing of their voices and do not have the community capital to create the green infrastructure that Louisiana has succeeded in developing.

Grassroots Coalitions and Environmental Efforts

In addition to the coalition on green infrastructure, several grassroots organizations have formed with the objective of working together to find innovative strategies to curate the change communities require. More than anything, however, people have stepped in and showed up for one another in ways their government never has or perhaps never could, given the ideological divide that consistently creates stalemate in the decision-making processes of the United States. Local mutual aid networks and grassroots organizations, social movements and community coalitions have soldiered on as they traditionally do. Even though some policies that are enacted are those that fail to address the legacy of racism that contributes to disaster in the first place, and politicians have failed to enact the same equitable protections and relief for communities of color that they do for wealthy, white communities, Black environmentalists in Louisiana are fighting for legislation to protect their communities from racist policies that have proven to worsen their social and environmental burdens. Community building, the decolonization of values and skill sets, coming back to nature and spirituality as a restorative and teachable space, and navigating political, social, and environmental disasters through solidarity are all ways in which Louisiana communities have dealt with the historical trauma induced by Western society, and ways of behaving that demonstrate resilience to it (Abate and Kronk Warner 2013). Smaller organizations have assumed roles to rebuild and replenish their communities where there is need for improved air quality, less soil degradation, gender equity, economic security, and environmental justice.

Parish Coalitions

RISE, in St. James Parish, Louisiana, is a Black, woman-led grassroots and faith-based organization that has protested Formosa Plastics in Cancer Alley. The group has used its faith and love to target elected officials as well as mobilize the community. During an incident of protest action where the community pressured its local officials to suspend the Formosa plastics permit, a leader in the organization referenced the biblical story of Zacchaeus, comparing the movement's officials to Zacchaeus who, though sinful and greedy, was found to be still worthy of love and grace and also capable of coming to terms with the truth and "coming down from the tree" (Soroski 2019). In Louisiana's faith-based communities, women often lead their community in song and prayer. On this occasion, they entreated (in a chant) their officials to "come down from the tree" and see the damage that Formosa was inflicting on their families' lives. Essentially, the community was advocating for local empowerment, inviting their public to be politically efficacious and their local government to be responsive to the needs of the people. Organizers convinced the community that every household had a stake in the fight to suspend Formosa's permit to build a gargantuan facility whose emissions would kill families in the neighborhood for generations to come.

RISE is only one of the local groups that has galvanized community opposition to the lack of government responsiveness to their environmental situation. Their activism has led to the successful defeat of action to construct a $1.25 billion plastics manufacturing plant in 2019 and the group is currently fighting to prevent Formosa Plastics from building a massive multibillion dollar plastics plant in the parish. It is this level of leadership, in identifying local knowledge capital, empowering those whose opinion leadership can bring the social capital into harmony to advocate for and empower stakeholders to demand public policy aimed at addressing the needs of vulnerable communities that this manuscript upholds. This evidence of democracy-in-action validates Dahl's (1989) argument that continuing responsiveness of a government to the preferences of its people must rely on the government's ability to perceive the preferences of the people. However, governments' perceptive ability requires the efficacy of community activists, who bring to light what needs to be done, what can be done, and how it needs to be done if indeed sustainable development can be achieved and publics can realize a quality life. Partnering with local government institutions, financial and technological agencies, and continuing to produce knowledge beneficial to small communities can jump-start development to a sustainable scale.

Concerned Citizens of St. John is another such organization. Chemical giant Dupont, now Denka, opened a plant in St. John's Parish, Louisiana in 1969, and it is the only facility in the country that manufactures neoprene. Chloroprene is a main ingredient in neoprene, and its emissions have contributed to five census tracts around the plant having the highest risk of cancer in the country, which is more than 700 times the national average (Hersher 2018). This facility was a former sugarcane plantation. Robert Taylor, eighty years old and founder of the Concerned Citizens of St. John, grew up as a remnant of this plantation system. Dupont moved in a year after he built a home for his family, an investment he thought, at the time, would benefit his family for generations. When industry moved in, no one was prepared. The impacts were devastating, and their lived environment changed drastically. Many succumbed to the devastating impacts of lifelong exposure to chloroprene emissions, and those still alive, battle debilitating diseases (Taylor 2022). When the EPA confirmed there was no place in St. John Parish with safe levels of chloroprene, Taylor, with the help of the Louisiana Environmental Action Network, founded the Concerned Citizens of St. John, a grassroots organization fighting to hold government officials accountable for the pollution and chemical emissions plaguing their Parish (Dermansky 2017).

The fight for self-determination is ongoing but the group continues to pressure the EPA to take emergency measures to curb the facility's emissions. Most recently, the group has sought help from the international human rights commission and filed an international human rights appeal over Denka's pollution in St. John. The last time this commission took substantial action in Louisiana was eleven years ago when it granted a hearing to Mossville, Louisiana, residents who charged that the US government violated their rights by not forcing local chemical plants to stop polluting their community (Baurick and Parker 2021). St. John is designated a "sacrifice zone." The Concerned Citizens of St. John hope that the United Nations can amplify their plight to the world and make clear that genocide is being perpetuated against the Black people of Cancer Alley, where the 92 percent of the population impacted by the petrochemical industry is Black, in a state where Black people are only 33 percent of the population (Taylor 2022).

As grassroots organizations have coalesced and advocated for rights relentlessly in the last decade, we note that without the sense of empowerment they experience from citizens standing together to demand their right to healthy air and water, public officials may not be willing to collaborate. The community's voice has the united power to make a difference to citizens' health and safety. Only through utilizing that voice both literally and figuratively will communities achieve the goal of self-determination and claim the sustainable future they have earned. Both RISE in St. James and the Concerned Citizens of St. John have a shared stake in their fight for clean air, water, and soil but they are not alone in this fight. Community, environmental, human rights, civil rights, and religious organizations have come together in solidarity to advocate for the health and safety of the river parishes in Cancer Alley (also known as Death Alley).

The Coalition Against Death Alley (CADA) announced a process of "nonviolent protests" to pressure their elected officials and petrochemical gargantuan facilities to stop poisoning their communities. This coalition has been relentless in their criticisms of the governor and other local officials and will continue to be until they protect the citizens they were elected to serve and give river parish communities a chance at "the good life" (Dermansky 2019). This coalition has successfully united communities up and down the Mississippi River: the poor, the disadvantaged, those designated expendable, Black and white, to fight for the good life they deserve. Together, they are the archers in their story, and their arrow hitting their target depends on their ability to set aside their differences (racial, religious, political, etc.), organize themselves, and create a united front to demand and create a livable future for the river parishes (Lee 2019).

Gulf Coast Alliances

The Gulf Coast Center for Law and Policy (GCCLP) is another social action group that demonstrates how much more can be achieved via community collaboration than passive tolerance of government inaction. As knowledge economy strategies suggest, locating knowledge capital at the grassroots level and producing that capital into innovative designs for environmental sustainability is what is needed to break free of the ceaseless wait on government funds and expertise to rebuild a community devastated by disaster. Although Louisiana is located in a developed country and has greater access to a just environment than people in the developing spaces of Latin America and the Caribbean, there is still much vulnerability in disaster-prone areas of the Black diaspora in the United States.

To address their needs, the Gulf South for a Green New Deal (GS4GND), under the umbrella of the GCCLP, has amassed a coalition of over 200 organizations advancing long-existing work in climate, economic, and racial justice across five states in the Gulf South: Louisiana, Texas, Alabama, Mississippi, and Florida. Launched in May 2019, this coalition is rooted in grassroots organizing led by frontline communities most impacted by disaster. They have advanced a uniquely southern vision and approach to the climate crisis and have grounded themselves and their fight in the strengths of their region: a region that plays a pivotal role in the US economy, national defense infrastructure, and the ongoing advancement of social innovation. Liberation work has always been a key part of Gulf South history, and this coalition has been successful in large part to this region's ability to retain such strong connections to culture, family, community, and land. They have leaned on the wisdom of generations of communities who have lived through the devastating impacts of climate change, environmental racism, and extractive industries, because they recognize frontline communities have unique and important perspectives for the attainment of a good life. They honor and center indigenous knowledge and wisdom in their work by healing and reconciling our human relationship to the land while engaging frontline communities in information-sharing, and participatory, collaborative, and intersectional approaches to liberation for all (Lux 2022). After the coalition launched, Colette Pichon-Battles, from the Gulf Coast Center for Law and Policy, facilitated efforts to author a policy platform that centered southern voices in the fight for a Green New Deal (Lux 2019). This living document started the conversation and gave the Gulf South a seat at the table to hold itself accountable to the unique requirements for a sustainable future (Lux 2022).

Along with law, policy, and other environmental coalitions, the Gulf of Mexico Alliance (GOMA), a regional partnership that focuses on enhancing the environmental and economic health of the Gulf via increased collaboration across the five Gulf states, has also been strident in its advocacy for post-disaster recovery and planning ahead for environmental sustainability. GOMA is part of a network of about 150 participating organizations from state and federal agencies, tribal governments, local communities, academia, non-governmental organizations, and industry. This network strives to empower communities, strengthen their resilience, share data, improve management of coastal habitats and wildlife species, and, very importantly, serve underrepresented groups in any way that would benefit their development. The Alliance believes that working together on common issues resulting from the region's disaster experiences is the best way to tap into local capital to find new and innovative solutions as a collective effort.

Since its inception in 2004, GOMA's achievements have been significant. The Alliance has succeeded in enlightening communities on issues leading up to disaster, post-disaster, and in preparation for further disaster. Through collaborative forums, the Alliance has developed and modified tools to address regional issues and enabled strategic partnerships. The organization has also been critical in tracking restoration efforts. It has earned a reputation for fostering relationships that lead to positive change and has given new meaning to collective energy, demonstrating how much more can be accomplished when groups work together than when a single entity attempts to create change. GOMA's efforts have been particularly outstanding for the diversity of its performance. It is seen as a trusted source of scientific, management, and policy information as a result of the wealth of intellectual capital of its partners and how seamlessly that capital can be produced to develop new ideas that take on the challenges of network communities with similar yet different needs for problem solving and solution identification. Empowered by the wisdoms of the 150 partners and more than a thousand members, the priorities of the region remain a focal point. The ocean partnership is inclusive and provides a commanding voice in the advocacy for a range of stakeholders and, in addition, the network provides coordination and leverage both public and private funding to realize common objectives. Facing the same challenges and finding creative ways to address those challenges allows the Alliance to anticipate next steps and prepare for any eventualities that may create distress for the Gulf. Calling themselves a regional ocean partnership means that they successfully tackle coastal and marine issues with equal efficacy and with solid support from local and regional communities. What GOMA has accomplished can be achieved by collaborative action among Caribbean and Latin American states, where equally underserved members of the Black diaspora abound. When collaborators listen to all stakeholders and learn from their feedback to redress negative output, they are actually engaging the essence of the knowledge economy model being advocated in this study.

Creative Community Project

The Walls Project, located in Baton Rouge, Louisiana, is a growing community organization that epitomizes local grassroots activism that fits well within the framework of knowledge economies, conceptualized as an alternative model for development in emerging societies seeking sustainable measures for growth. The organization's mission is to stimulate Louisiana's economy via creative pathways. This organization envisions performance in three primary areas:

- creativity by way of painting murals in underinvested schools and neighborhoods, which now serve to motivate communities to honor their past and incorporate it into their present to prepare youth for leadership in the future.
- cultivation of knowledge production by educating and inspiring youths to attain jobs in high demand but for which too few from the community are qualified to participate.
- reactivation of communities by remediating blight and making communities safer and healthier.

In further evidence that Louisiana recognizes the path to development is through the empowerment of its human capital, the Walls Project perceives local communities to be the appropriate change-makers and takes on the role of enlightening others and "challenging preconceptions" in order to unite individual members of the community by "celebrating their differences and igniting progress through dialogue and action" (Williams 2021). In an address to a panel on the Innovation Station, organized by the State Department's Global Women's Issues desk and moderated by Aubrey Paris, Helena Williams outlined the mission of the Walls Project as one envisioning a vibrant, creative economy, accessible to all. The project stresses its areas of focus as public art, workforce training, and blight remediation. This, the project promoters understand as having the capacity to overcome poverty while empowering communities to build culture and be inclusive. The project is funded by generous donations from nonprofits and individual contributors but realization of its vision comes entirely from the community as a bottom-up, rather than the usual top-down, venture for investment.

What the project has brought to the community has been aptly matched by the energies the community invests in the project's development. The idea of the project is not limited to the minorities in Louisiana's capital city but has seen expansion in communities such as Dallas, Texas, and could equally be extended to areas of the diaspora in the less developed and mostly diasporic community spaces in the Americas and Caribbean. Walls have traditionally been built to enclose and protect similarities or exclude dissimilarities. The Walls Project in Louisiana has successfully led programs and events that have built alliances and torn down limiting societal walls that discourage safe, healthy living and prosperous lives. Over the last decade, this project has contributed to economic growth in the area of approximately $3 million, in partnership with nonprofit organizations, government agencies, and empowered communities. Their progress in the area of public art, urban agriculture, and community reactivation stands as a model for the Black diaspora located outside of Global North sites where a similar need for progress in the molding of human capital exists. There is no doubt that with knowledge-sharing across the Black diaspora, we might see the development gap closing and knowledge economies rising to empower future analysts, problem-solvers, and policy-makers. Movement away from the individualism of the developed world to the collective power that must be harnessed in emerging economies, if progress is to be achieved, is the hope of the future for the far-flung Black diaspora, whose urgency to mitigate disaster risks and prepare for a green, carbon-limited, safe, and healthy world cannot be delayed.

Summary

In Louisiana, disaster has presented itself in numerous forms: super storms, flooding, crumbling infrastructure, government negligence, pandemics, unemployment, corporate greed, and extractive industries, just to name a few. Some would say that Louisiana is both vulnerable and resilient to disaster but these disasters are man-made, and they cannot ever be detached from the context of colonialism, which created both the economic conditions for disaster and the social conditions that limit our capacity to achieve the full dimensions of resiliency. Over the last decade, Louisiana's production of local knowledge and its advocacy for and empowerment of community partnerships have resonated throughout the entire Gulf Coast and are preparing the way for a remarkable turnaround in fortunes as its collaborative energies are engaged in the fight against the onslaught of climate change.

The examples included in this chapter illustrate how communities in Louisiana have acted locally, leaning into their traditional voices and teachings to mobilize their communities, while embracing the diverse livelihoods, cultural values, and social networks that contribute to their adaptive capacity to survive and resist disaster. Advancing grassroots policy and practices that center Black and brown people, laborers, farmers, fisher-folk, tribal nations, and frontline communities (see figure 3.1 below) in a just transition away from extractive economies is tantamount to determining the very survival of the state (First Peoples Conservation Council 2022). But how much more must our communities do to secure the kind of sustainable outcomes necessary for survival, and is survival alone

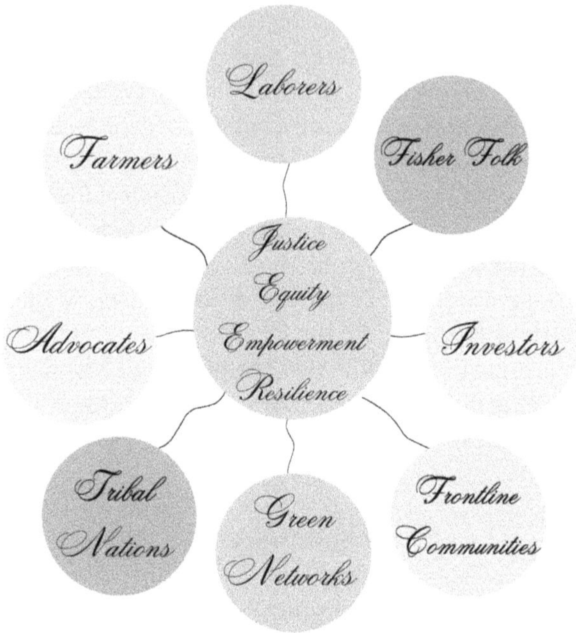

Figure 3.1. Mapping Engaged Louisiana Community. © Pamela Waldron-Moore 2022.

enough? Our traditional knowledge is fundamental to understanding our lived environment, resistance activities, and the resilience and adaptation strategies we adopt that inspire ingenuity. With this we have the ability to solve pressing life challenges and prevent disaster by situating our solutions from within our own deep spatial knowledge. Local indigenous and diasporic knowledge systems and practices are a major resource for striving toward the "good life" and integrating these forms of knowledge with existing practices, increasing our capacity to acquire it. Differences between traditional and Western worldviews mean that incorporating the two can be difficult. Some argue that merging them is not possible, but rather one must recognize that the awareness of each way of knowing can inform new solutions and provide us with the opportunity and capacity to hit our target and live a quality life (Alexander et al. 2011).

How we manage and strive toward our target has much to do with whom it impacts, what their experience is, where it is located, and what resources are available to attain our goals. Are there strong crosswinds? If so, we must aim slightly into the wind. If we aim into the wind, we must aim high to compensate for its slowing effect. Our concept of the "good life" continually changes in our journey to pursue it (Koehn 2012).

Our sense of truly equitable development, as it relates to our own lived environment and ever evolving conditions, also changes throughout our journey to pursue it. It requires that we rebuild with new tools, born of our knowledge capital and produced for our long-term development. Figuring out what makes a human life good has implications for how we treat those whose lives are different from our own. Disasters comes in all shapes and sizes as does their impact on communities and countries of all shapes and sizes. Movements, coalitions, and knowledge-sharing to scale have long been united to create a resilient and sustainable future for all, and our collective fight is anchored in the histories, realities, and power of this region and others like Rwanda and Haiti. It is the only way we can adopt the sustainable development approaches necessary for our survival and attainment of the good life. We must pool our collaborative energies in pursuit of the environmental sustainability our region requires to succeed beyond the level of mere survival. We must deputize ourselves, rather than wait for a government's permission, to produce new knowledge to mitigate the risks of climate change and the attendant disasters that impact our vulnerable communities.

Jessica Dandridge is the Executive Director of The Water Collaborative of Greater New Orleans. She has worked in community advocacy, campaign development, and strategic planning for NGOs for seventeen years. Ms. Dandridge received her BA at Xavier University of Louisiana in Political Science, and her MA in International Affairs with a concentration in conflict and security at The New School for Public Engagement in New York City. Currently, Jessica advises federal agencies as the Co-Chair of the National Academy of Science Resiliency Roundtable and is a Commissioner for the Governor's Advisory Commission on Coastal Protection, Restoration, and Conservation.

Yeishka Montalvo graduated with a BA in Political Science from Xavier University of Louisiana in 2019. Based in Louisiana, Yeishka serves as the Latine Constituency Manager for the Sunrise Movement—a movement of young people pushing to make climate change an urgent priority and build toward a livable future. Disaster became a live interest for her after Hurricane Maria ravaged Puerto Rico in 2017. Through the pain of disasters hurting the people and places she loves, she witnessed the power of her community to fight for dignity and justice with creativity and resolve, and committed herself to environmental activism.

References

Abate, Randall S., and Elizabeth Ann Kronk Warner. 2013. "Commonality among Unique Indigenous Communities: An Introduction to Climate Change and Its Impacts on Indigenous Peoples." In *Climate Change and Indigenous Peoples: The Search for Legal Remedies*, ed. Randall S Abate and Elizabeth Ann Kronk Warner, 3–18. Northampton, MA: Edward Elgar Publishing.

Adams, Vincanne, Taslim Van Hattum, and Diana English. 2009. "Chronic Disaster Syndrome: Displacement, Disaster Capitalism, and the Eviction of the Poor from New Orleans." *American Ethnologist* 36(4): 615–36.

Alexander, C., N. Bynum, E. Johnson, U. King, T. Mustonen, P. Neofotis, N. Oettlé, C. Rosenzweig, C. Sakakibara, V. Shadrin, M. Vicarelli, J. Waterhouse, and B. Weeks. 2011. "Linking Indigenous and Scientific Knowledge of Climate Change." *BioScience*. 61(6): 477–84.

Baurick, Tristan, and Halle Parker. 2021. "It Won't Stop Raining in New Orleans Area: Experts Explain Why It's Been Brutal." *The Times Picayune/Nola.com*, 20 July. Retrieved 23 April 2023 from https://www.nola.com/news/environment/article_4a07adfa-e97f-11eb-ac34-6f67ec6302a6.html.

Bay Area News Group. 2016. "A Third World Nation within Our Own." *East Bay Times*, 17 August. Retrieved 23 April 2023 from https://www.eastbaytimes.com/2005/09/04/a-third-world-nation-within-our-own-2/.

Bill of Rights Institute. n.d. "The Trail of Tears." *Bill of Rights Institute*. Retrieved 23 April 2023 from https://billofrightsinstitute.org/essays/the-trail-of-tears.

Blodgett, Abigail. 2006. "An Analysis of Pollution and Community Advocacy in 'Cancer Alley': Setting an Example for the Environmental Justice Movement in St. James Parish, Louisiana." *The International Journal of Justice and Sustainability* 11(6): 647–61.

Bradshaw, Jim. 2011. "Great Flood of 1927." *64 Parishes*, 13 May. Retrieved 23 April 2023 from https://64parishes.org/entry/great-flood-of-1927.

Bridges, Todd S., E. Michelle Bourne, Burton C. Suedel, Emily B. Moynihan. 2021. *Engineering with Nature: An Atlas*. Vol. 2. Vicksburg, MS: U.S. Army Engineer Research and Development Center, Environmental Laboratory and Information Technology Laboratory.

Broach, Drew. 2022. "Hurricane Ida Makes Top 10 for Costliest Storms in U.S History." *The Times Picayune/Nola.com*, 6 February. Retrieved 23 April 2023 from https://www.nola.com/news/hurricane/article_716be8ee-83cf-11ec-a605-a7749397ceec.html.

Bruneau, Michel, Stephanie Chang, Ronald Eguchi, George Lee, Thomas O'Rourke, Andrei Reinhorn, Masanobu Shinozuka, Kathleen Tierney, William Wallace, and Detlof Von Winterfeldt. 2003. "A Framework to Quantitatively Assess and Enhance the Seismic Resilience of Communities." *Earthquake Spectra* 19(4): 733–52. https://doi.org/10.1193/1.1623497.

Burdreau, Cain. 2018. "Oil Canals' Role in La. Land Losses Debated." *ABC News*. Retrieved 23 April 2023 from https://abcnews.go.com/Business/story?id=4162431&page=1.

Burton, O. Vernon, Troy Smith, and Simon Appleford. n.d. "Forced Over the Great River: Native Americans in the Mississippi River Valley, 1851–1900." *Northern Illinois University Digital Library*. Retrieved 23 April 2023 from https://digital.lib.niu.edu/twain/forced.

Campanella, Richard. 2016. "Beneath New Orleans, A Coastal Barrier Island." *64 Parishes*, 1 December. Retrieved 23 April 2023 from https://64parishes.org/46028-2.

Castellón, Idna. 2021. "Cancer Alley and the Fight Against Environmental Racism." *Villanova Law: Environmental Law Journal* 32(1). Retrieved 23 April 2023 from https://digitalcommons.law.villanova.edu/elj/vol32/iss1/2.

Dahl, Robert. 1989. *Democracy and Its Critics*. New Haven, CT: Yale University Press

Davis, John Dean. 2018. "Levees, Slavery, and Maintenance." *Society of History and Technology*, 20 August. Retrieved 23 April 2022 from https://www.technologystories.org/levees-slavery-and-maintenance/.

De Cesar, Wayne T., and Susan Page. 2003. "Jefferson Buys Louisiana Territory and the Nation Moves Westward." *Prologue* 35(1). Retrieved 23 April 2022 from https://www.archives.gov/publications/prologue/2003/spring/louisiana-purchase.html.

Dermansky, Julie. 2017. "Louisiana's Cancer Alley Residents Sue Chemical Plant for Nearly 50 Years of Air Pollution." *Desmog*. https://www.desmog.com/2017/07/27/louisiana-cancer-alley-sue-denka-dupont-chemical-plant-50-years-air-pollution/.

———. 2019. "Outraged, New Coalition Emerges against Louisiana's Expanding—and Polluting—Petrochemical Industry". *Desmog*, 28 March. Retrieved 23 April 2023 from https://www.desmog.com/2019/03/28/cancer-coalition-against-death-alley-louisiana-expanding-petrochemical-industry/.

"Dutch Dialogues: New Orleans Talks Resiliency with the Netherlands." 2014. *US Green Building Council*, 7 October. Retrieved 23 April 2023 from https://www.usgbc.org/articles/dutch-dialogues-new-orleans-talks-resiliency-netherlands.

First People's Conservation Council (website). 2022. Retrieved 23 April 2023 from https://fpcclouisiana.org.

"Fiscal Year (FY) 2014 Funds National Disaster Resilience Competition." 2014. "MRGO: Mississippi River Gulf Outlet." n.d. *Pontchartrain Conservancy*. Retrieved 23 April 2023 from https://scienceforourcoast.org/pc-programs/coastal/coastal-projects/mrgo-mississippi-river-gulf-outlet/.

Flowers, Catherine Coleman. 2021. "Hurricane Ida Shows the One-Two Punch of Poverty and Climate Change." *Nature*, 21 September. Retrieved 23 April 2024 from https://www.nature.com/articles/d41586-021-02520-8.

Forrest, Sharita. 2019. "Study Examines Effects of Climate Change, Land Loss on Louisiana's Houma Tribe." *Illinois News Bureau*, 27 September. Retrieved 23 April 2023 from https://news.illinois.edu/view/6367/803055.

Gabour, Jim. 2017. "New Orleans under Water: 12 Years after Katrina, Officials Can't Get It Right." *The Guardian*, 15 August. Retrieved 23 April 2023 from https://www.theguardian.com/us-news/2017/aug/15/new-orleans-flooding-rain-water-louisiana.

Garrigus, John D., and Gwendolyn Midlo Hall. 1994. "Africans in Colonial Louisiana: The Development of Afro-Creole Culture in the Eighteenth Century." *The William and Mary Quarterly* 51 (3). JSTOR: 547

Gerring, John. 2010. *A Centripetal Theory of Democratic Governance*. Boston: Cambridge University Press.

Greater New Orleans Foundation. 2021. *Economic Opportunity through Green Infrastructure: Equitable Access to Jobs and Contracts*. New Orleans: Center for Philanthropy. Retrieved 23 April 2023 from https://issuu.com/greaterneworleansfoundation/docs/gnof20_green_final_digital-2020-12-23.

Hannah-Jones, Nikole, Sean Wilenz, and Jake Silverstein. 2019. *The 1619 Project*. NY: The New York Times Magazine, 4 September. Retrieved 23 April 2023 from https://www.nytimes.com/interactive/2019/08/14/magazine/1619-america-slavery.html.

Hersher, Rebecca. 2018. "After Decades of Air Pollution, a Louisiana Town Rebels against a Chemical Giant." *NPR*, 6 March. Retrieved 23 April 2023 from https://www.npr.org/sections/healthshots/2018/03/06/583973428/after-decades-of-air-pollution-a-louisiana-town-rebels-against-a-chemical-giant.

Jacobs, David. 2023. "See why Louisiana Ranks last on US News Best States." Retrieved 23 April from https://www.businessreport.com.

Johnson, Fry & Co. 1858. "DeSoto's Discovery of the Mississippi, 1541." *Gilder Lehrman Collection*. Retrieved 23 April 2023 from https://www.gilderlehrman.org/history-resources/spotlight-primary-source/de-sotos-discovery-mississippi-1541.

Koehn, Glen. 2012. "The Archer and Aristotle's Doctrine of the Mean." *Peitho Examina Antiqua* 3(1): 155–68. https://doi.org/10.14746/pea.2012.1.7.

Lee, Bob. 2019. "Fighting to Survive in Louisiana's Death Alley: Interview with Death Alley Activists." *People's Tribune*. Retrieved 23 April 2023 from http://www.peoplestribune.org/latest-news/fighting-to-survive-in-louisianas-death-alley.

Levenson, L., 2018. "New Orleans, 300, Bulbancha, 3000." *Tripod*. https://tripodnola.org/episodes/new-orleans-300-bulbancha-3000/.

Louisiana Rankings. 2022. US News Best States. https://www.usnews.com.

Louisiana State Museum. n.d. "The Cabildo: Two Centuries of Louisiana History; Antebellum Louisiana II: Agrarian Life." Retrieved 23 April 2023 from https://www.crt.state.la.us/louisiana-state-museum/online-exhibits/the-cabildo/antebellum-louisiana-agrarian-life/index.

Luedke, Heather. 2019. "Fact Sheet: Nature as Resilient Infrastructure – An Overview of Nature-Based Solutions." EESI, 16 October. Retrieved 23 April 2023 from https://www.eesi.org/papers/view/fact-sheet-nature-as-resilient-infrastructure-an-overview-of-nature-based-solutions.

Lux, Travis. 2018. "NOLA vs Nature: The Blessing and Curse of the Wood Screw Pump." Tripod: New Orleans at 300. *WWNO*, 11 January. Retrieved 23 April 2023 from https://www.wwno.org/podcast/tripod-new-orleans-at-300/2018-01-11/nola-vs-nature-the-blessing-and-curse-of-the-wood-screw-pump.

———. 2019. "Advocates Launch Green New Deal Platform for Gulf South." *WWNO*, 14 November. Retrieved 23 April 2023 from https://www.wwno.org/coastal-desk/2019-11-14/advocates-launch-green-new-deal-platform-for-gulf-south.

Lyer, Deepa. 2022. *Social Change Now: A Guide for Reflection and Connection.* Washington, DC: Thick Press.

Mahoney, Adam. 2021. "'It doesn't have to be this way': Lessons from the slow death of Louisiana's oil industry." *Grist.* Retrieved 23 April 2023 from https://grist.org/climate-energy/it-doesnt-have-to-be-this-way-lessons-from-the-slow-death-of-louisianas-oil-industry/

Meehan, Miranda, Kevin Sedivec, Thomas DeSutter, Chris Augustin, and Aaron Daigh. 2017. "Environmental Impacts of Brine (Produced Water)." *North Dakota State University* R1850. Retrieved 23 April 2023 from https://www.ag.ndsu.edu/publications/environment-natural-resources/environmental-impacts-of-brine-produced-water/r1850.pdf.

Mendie, John Gabriel, and Stephen Nwanaokuo Udofia. 2020. "The Idea of the Good Life in Aristotle and Confucius." *Jurnal Sosialisasi: Jurnal Hasil Pemikiran, Penelitian Dan Pengembangan Keilmuan Sosiologi Pendidikan* 7(1): 61–73. https://doi.org/10.26858/sosialisasi.v0i1.14529.

Monteverde, Danny. 2021. "Thousands of Blighted Properties in New Orleans Create Problems for Neighbors." *Eyewitness News*, May 5. Retrieved 23 April 2023 from https://www.wwltv.com/article/news/local/orleans/thousands-of-blighted-prop

erties-in-new-orleans-create-problems-for-neighbors/289-a0823e0e-98d9-4601-945d-f9ea7d442ed0.

Ouchley, Kelby. 2015. "Hurricane Betsy." *64 Parishes*, 10 September. Retrieved 23 April 2023 from https://64parishes.org/entry/hurricane-betsy.

"Overview of Louisiana." 2021. *USNews*, 9 March. Retrieved 23 April 2023 from https://www.usnews.com/news/best-states/louisiana.

Parrish, Susan S. 2017. "The Great Mississippi Flood of 1927 Laid Bare the Divide between the North and the South." *Smithsonian Magazine*, 11 April. Retrieved 23 April 2023 from https://www.smithsonianmag.com/history/devastating-mississippi-river-flood-uprooted-americas-faith-progress-180962856/.

Perkins, Emily, and John Magill. 2020. "In the Late 1800s, Devastating Yellow Fever Epidemics Forced New Orleans to Confront Its Sanitation Problem." *The Historic New Orleans Collection*, 12 May. Retrieved 23 April 2023 from https://www.hnoc.org/publications/first-draft/late-1800s-devastating-yellow-fever-epidemics-forced-new-orleans-confront.

Phillips, Daniel. 2020. "Fading Away: Louisiana's Battle against Coastal Erosion." *KATC: Acadiana's Newschannel*, 23 November. Retrieved 23 April 2023 from https://www.katc.com/news/fading-away-louisianas-battle-against-coastal-erosion.

Powell, Lawrence N. 2013. *The Accidental City*. Boston: Harvard University Press.

Rodin, Judith. 2015. "A Roadmap to Resilience/New Orleans: Ten Years Later." *YouTube*, uploaded by AtlanticLIVE on 25 August. Retrieved 23 April 2023 from https://www.youtube.com/watch?v=PBZgzfLpyKs.

Samuels, Diana. 2017. "Looking Back at the Louisiana Flood of 2016: From 2 Feet of Rain to Sodden Drywall." *The Times Picayune/Nola.com*, 12 August. Retrieved 22 April 2023 from https://www.nola.com/news/weather/article_843a2f36-b5ff-5f06-b361-0c328b09c9e9.html.

Schleifstein, Mark. 2022. "Louisiana's Coastal Communities Could See $5.5B in Damages in Worst-Case Climate Change Scenario." *The Times Picayune/Nola.com*, 25 April. Retrieved 22 April 2023 from https://www.nola.com/news/environment/louisianas-coastal-communities-could-see-5-5b-in-damages-in-worst-case-climate-change-scenario/article_5c828372-c25c-11ec-96ee-83a2a25ac3de.html.

Severin, Giles Timothy, Robert A. Muller, Richard H. Kesel, Randall J. Schaetzel, and the Editors of Encyclopaedia Britannica. n.d. "Mississippi River." *Encyclopaedia Britannica*. Retrieved 23 April 2023 from https://www.britannica.com/place/Mississippi-River/Physical-features.

Soroski, Jason. 2019. "Who Was Zacchaeus? 5 Ways We Can All Relate to His Story in the Bible." *Crosswalk*, 10 April. Retrieved 23 April 2023 from https://www.crosswalk.com/faith/bible-study/5-ways-we-can-all-relate-to-zacchaeus.html.

Taylor, Robert. 2022. "We Have Been Designated a Sacrifice Zone." *Poor People's Campaign*. Retrieved 23 April 2023 from https://www.poorpeoplescampaign.org/we-cry-power/robert-taylor/.

Turmaud, Danielle Render. 2020. "What Is Collective Trauma? How It Could Be Affecting Us." *Psychology Today*, 23 May. Retrieved 23 April 2023 from https://www.psychologytoday.com/us/blog/lifting-the-veil-trauma/202005/what-is-collective-trauma.

US Department of Housing and Urban Development (HUD), 17 September. Retrieved 23 April 2023 from https://www.hud.gov/program_offices/administration/grants/fundsavail/nofa14/ndrc.

US EPA. 2022. "Benefits of Green Infrastructure." *US Environmental Protection Agency*, 7 September. Retrieved 23 April 2023 from https://www.epa.gov/green-infrastructure/benefits-green-infrastructure.

Waggonner & Ball. n.d. [2014] "Greater New Orleans Urban Water Plan." Waggonner & Ball. Retrieved 23 April 2023 from https://wbae.com/projects/greater-new-orleans-urban-water-plan-2/.

Wells, B., and K. Wells. 2005. "First Louisiana Oil Wells." *American Oil & Gas Historical Society*, 1 September. Retrieved 23 April 2023 from https://aoghs.org/petroleum-pioneers/first-louisiana-oil-well/.

Wilburn, Brad. 2015. "Spice for the Good Life." Markkula Center for Applied Ethics at Santa Clara University, 12 November. Retrieved 23 April 2023 from https://www.scu.edu/ethics/ethics-resources/ethical-decision-making/spice-for-the-good-life/.

"Yellow Fever." 2022. *Centers for Disease Control and Prevention*, 22 September. Retrieved 23 April 2023 from https://wwwnc.cdc.gov/travel/diseases/yellow-fever.

CHAPTER 4

Haiti

Building Social Capital through Media Connectivity

SHEARON ROBERTS

Introduction

The public sphere is central to the collective shaping of local ways of knowing. Through citizen engagement in dialogue, debates, advocacy and even creative forms of expression, societies propose and refine notions of ways of life and uphold commonly shared values of humanity and dignity. Media platforms, both traditional and digital, offer forms of public spheres that transmit public knowledge, stimulate debates on the status quo, and serve as a rallying call for action in the face of persistent political and social stalemates. In Haiti, traditional media, particularly commercial and community radio, played an important role during the 2010 earthquake disaster and the decade since in Haiti's quest for a more equitable recovery and sustainable way of life for ordinary Haitians. Since 2018, digital platforms have bolstered local and global conversations that aim to hold those in power accountable for unsustainable forms of governance that continue to increase hardships for Haitians. This chapter examines the decade since the 2010 earthquake in Haiti, emphasizing the ways the media serves as a public sphere and a civic institution that generates, disseminates, and curates a local knowledge economy in search of equitable post-disaster recovery that benefits all Haitians. Further, this chapter offers a perspective of the ways in which knowledge economies can be developed through the public sphere in ways that include all citizens and not just those in power, or social and economic elites alone.

The Catastrophe of 2010

In less than twenty-four hours after a 7.3 magnitude earthquake shook Port-au-Prince, the capital of Haiti, on 12 January 2010, citizen brigades arrived from rural and southern parts of the country, offering food, supplies, tools, and even their hands to pick through the rubble to find the living and to aid in their survival (Bojarskie, André, and Pierre-Pierre 2020; Schuller and Morales 2012). One-third of all Haitians were impacted by this disaster, one the country had not experienced since the 1700s. More than 200,000 Haitians lost their lives.

On 4 October 2016, when Hurricane Matthew, a Category 4 storm, made landfall in the southwest of Haiti, it destroyed over 90 percent of crops and homes in rural provinces. Haitians who lived in the capital of Port-au-Prince, which was slowly rebuilding, emerged to coordinate the relief of their fellow citizens who had lost all they had, bringing immediate food and supplies, and aiding in temporary shelter. Urban Haitians had, in essence, returned the benevolence they had received from rural Haitians six years prior when faced with another disaster. Over 500 Haitians died in the mid-decade disaster, and thousands were forced to evacuate and were displaced from their homes for several months. Meanwhile, public services, schooling, economic activity, and agriculture all came to a grinding stop in the southwest region.

Two years later, Haitians embarked on an exercise of sustained civil engagement in their public sphere that intensified from 2018 to 2021. To outsiders, it was described as "civil unrest." The descriptive "unrest" places emphasis on the disruption, vandalism, and even violence that comes when citizens take to the street in search of institutional accountability. What a term like "unrest" often biases is the diminution of and shifts in attention from the very act by citizens of shaking the status quo in ways that call attention to inequities that can no longer be sustained among a populace. In the Haiti context, this is a citizenry that had endured the vulnerability of natural disasters that can be traced back to the first decade of the 2000s and became increasingly harder to recover from after 2010 and 2016.

The source of Haitians taking action from 2018 to 2021 was simple: knowledge. Haitian citizens had acquired knowledge on why its country had failed to fully recover from disasters and Haitians were seeking knowledge on how to stabilize their country and to protect citizens from the natural threats to come. It started with a question: "Kot Kòb PetroCaribe a?" (Where is the PetroCaribe money?) (Mirambeau 2018). It was answered with the hashtag: #PetroCaribeChallenge in 2018, a digitally networked movement that was as much offline as it was online.

The next year, as illustrated in the figure below, Haitians mobilized under *peyi-lòk* (country shutdown). Schools and public transportation services shut down in 2019 as Haitians called for government accountability for the billions of dollars directed to aid citizens and the nation as a whole to fully recover from 2010 and 2016.

Figure 4.1. #PetwoKaribe Movement. Mobilizing Social Capital Both Online and Offline (Mirambeau 2018) | #PetroKaribe Movement. Educating and Raising Awareness on Post-Disaster Sustainable Recovery (IJDH 2018).

In 2020, Haitians sustained their calls for accountability, as the COVID-19 pandemic exacerbated hardships for citizens. They particularly aimed to hold up their constitution, calling for leadership change in 2020 and for elections that would usher in leaders who would work to ease economic distress and begin the hard work of addressing inequity.

On 7 July 2021, Haiti's president Jovenel Moïse was assassinated at his home and his wife Martine Moïse sustained injuries in the attack. A year later, Haitians continued to hold the appointed acting Prime Minister Ariel Henry accountable on when elections can be held so that the work of rebuilding the country can truly commence.

Whereas Haiti faces challenges in human and economic capital, it is a beacon in the arena of social capital. Social capital's value rests in the investment of people, or citizens, in forming networks that can deliver outcomes and pressure a society to function optimally or to enable change for the effective functioning of a society. Theorists of social capital posit that it is a key resource in aiding societies to harness human capital and to build up and sustain economies.

Civil society coordinates social capital through networks that can be communal, institutional, cultural, faith-based, or otherwise. Those networks are fostered through the linkages and ties of groups in which knowledge and resources are shared and exchanged. When networks come together in larger societal movements, they are bolstered by the convening of the public sphere. History shows that the coordination of movements is often supported by traditional forms of media. In Haiti in particular, the press has historically played a key role in the shaping of the world's first Black republic (Stieber 2020). At the turn of the century, it advocated for the country's sovereignty as foreign powers encroached on Haiti, resulting in the US invasion and occupation from 1915 to 1934 (Roberts 2016). Radio in Haiti then emerged as a post-occupation vehicle for educating and coordinating the public sphere as Haiti endured several decades of dictatorship, supported and enabled by foreign powers (Wagner 2017).

From 2010 to 2021, Haiti expanded its digital landscape. Digital media nationalized and globalized the relay and transfer of information and knowledge that aided Haitians both in recovering from disasters in citizen-centered ways, but also in mounting a national exercise to dismantle institutions that do not work for citizens and to call for reforms and changes that can build back a nation that is still to recover from the disasters it has seen and the ones that loom in the future.

This chapter explores how mediated knowledge dissemination can aid in citizen-centered post-disaster recovery. It also maps how the expansion of digital media coincides with a widening of the public sphere, contemporizing the formation of social capital networks that work to sustain movements that aim to advocate for the full and effective functioning of societies, when a fledgling democracy disenfranchises its citizens and fails to work to alleviate the hardships and inequalities its citizens must endure.

Social Capital and Dissemination of Knowledge within the Public Sphere

When we think of the resources within societies, social capital often holds the most potential for the dissemination of knowledge. Eric Lesser (2000) points to two key distinctions of social capital: it is a "public good" and it is "'located' not in the actors themselves but in their relations with other actors" (Lesser 2000: 8). Human or physical capital requires "individual" or "collective ownership," but since social capital inherently seeks the public good, it cannot be owned. Additionally, social capital is mobilizing and capacity-building. Lesser writes that it "cannot exist in a vacuum, it is dependent on the interaction of individuals to create value" (2000: 8). Social capital's value rests in its future returns, its ability to be converted into other forms of capital, its linking and connectivity of society around a common goal, and compared to physical capital, it cannot depreciate.

Lesser explains that "social capital, much like organizational knowledge, often grows and becomes more productive with use" (2000: 8). The more societies activate and invest in its social capital, the more social capital works for the good of societies.

Social capital relates to knowledge in that "social capital is necessary to enable the effective management of both explicit and tacit knowledge" (Lesser 2000: 9). It works as such: the more individuals in a system trust those that deposit knowledge into the system, the more that knowledge will be circulated or disseminated. The more individuals in a system see value in knowledge, the more likely they are to contribute and disperse that knowledge. The more individuals can identify and share commonalities with members within a system, the more likely they are to pass on that knowledge. Social capital enables and bolsters the active engagement with and use of knowledge, contextualizing it for networks, and allowing a networked society to convert that capital into other forms of capital. As a key theorist on the networked economy, Robert Putnam (1993) notes that civic participation in local, communal affairs is central to social capital (1993: 37). In societies with strong social capital, civic participation has been built up over many years. The coordination of civic participation is often mostly studied in Western societies, and what that civic participation often looks like is usually based on Western norms. Yet, Haitians have placed their own cultural stamp on participation, as Figure 4.1 above clearly illustrates.

Since the founding of Haiti as an independent republic, civic participation through its networked economy has been at the foundation of building a state that the Western world was hostile to. But Haiti also problematizes the traditional thinking of the convertibility of social capital, as theorists have applied it to date. Putnam indicates that "the social capital embodied in the norms and networks of civic engagement seems to be a precondition for economic development as well as for effective government" (Lesser 2000: 11; Putnam 1993: 37). In studying networks in Europe, scholars like Putnam note that networks enabled communities to pool labor and resources, transferring knowledge to improve local economies and bolstered distressed industries after a downturn. Putman notes that "trust lubricates social life" (Lesser 2000: 11; Putnam 1993: 37). Applications outside of Western case studies indicate that Haiti does not at first glance fit the Western test case on social capital. Certainly Haiti's social capital is rooted in strong and enduring civic networks that carry trust and shared norms and is rooted in histories of civic participation, but where Haiti diverges as a case study of social capital is in how its value is ultimately converted, automatically, to other forms of capital. For, while civil society has shared ownership in Haiti of its social capital, it cannot translate its vibrant social capital to other capitals because Haiti today does not "own" its economy or its governmental authority.

Since the 2010 earthquake, Haiti's economy is dependent on non-governmental agencies and multilateral agencies. Its elections have been either upheld or determined by international bodies or governments. Indeed, both US presidents Donald Trump and Joe Biden agreed that Haiti's President Moïse could delay the country's election and remain in power, in violation of its constitution, despite four years of civil society calling for free and fair elections, government accountability, and governmental change. If nothing else, this external support demonstrated to Haitians that the interests of Haiti's public were of little value to the developed world and continued to mirror the benign neglect of Haiti's wealthiest neighbors.

Thus, the case of Haiti calls us to expand our understanding of the value of social capital. Where scholars have often assumed a linear trajectory from active civic participation to effective economies and governance, Haiti prompts us to consider that Haiti's path is curvilinear, with twists and curves in theorizing how social capital gets converted to other forms of capital. It underscores even more in the Haiti case the importance of social capital because its persistence and endurance when societies are disenfranchised economically, socially, and politically is what must be amplified to truly achieve a functioning society when other institutions fail. Thus, the amplification of social capital is often a function of the public sphere. When the public sphere co-opts and coordinates with media, it intensifies the transfer of knowledge, it manages that transfer of knowledge, and it sets a course for how that knowledge can move societies out of prolonged crises and into sustainable recovery. In examining Haiti, this factor is crucial because civil society requires other forms of power to effect change in the spaces it lacks power. The media and its ability to democratize power, by elevating citizens and networks through platforms, bolster the work of social capital and allow networked societies to leverage that mediated power in pursuit of the public good.

The Contemporary Mediated Public Sphere

Social capital theorists note that technology that can support the movement of knowledge derived from social capital is both a form of management of the dissemination of knowledge and a tool for the dissemination of knowledge. The dominant technology today utilized by networks in society is our digital media landscape. This spans our traditional media platforms now operating online, our new media platforms, social media, mobile applications, and remote technologies. Prior to our digital landscape, the media often served as a Fourth Estate, a check and balance on other forms of institutional power in a society, be it social, economic, or political, regulating and providing balance to institutional powers in maintenance of the public good (Lawson 2002).

However, media power until the digital revolution was primarily controlled in many societies. Control of the press by families or corporations often concentrated power into a small elite hand that could control the message, persuade governments, and control the masses (Hughes and Lawson 2005; Lugo 2008). Concentrated media power can also marginalize and exclude portions of society. In the United States, mainstream media excluded for centuries the plight of African Americans from their enslavement up to the present day and the systems put in place to devalue their full citizenship as Americans (Chafee 1947). In Haiti, mainstream media until the proliferation of radio broadcasting widely marginalized the masses, reporting in French, the language of elites and colonizers, to the exclusion of peasants who could only transfer information and knowledge in the language spoken by all: Kreyol (Montas-Dominique 2002).

The digital revolution disrupted media control (Jenkins and Thorburn 2003). It democratized the media and opened access to those who can create, curate, and disseminate mediated messages. It also globalized the sharing, management and contextualizing of knowledge across societies. Within the last two decades, the digital revolution has allowed for global learning to take place across continents and in remote locations. It has also allowed social networks to connect across nations, removing borders and boundaries, raising solidarity, and facilitating transnational movements (Hudson and Zimmerman 2015; Jorba and Bimber 2012; Juris 2005) without the need for agents or activists to ever get on an airplane.

Certainly, not all cost barriers have been removed because of the digital age. While citizens no longer need to purchase or own a printing press, a broadcast station, a transmitter, or a license to gain access to a media platform, they would need access to broadband, the internet, a computer, or mobile device. Depending on the scale of information gathering, creation, and dissemination, other costs accumulate in sustaining the transfer of knowledge over new media platforms. Additionally, new media platforms are often controlled and regulated by the private companies that own them, and they have executed that control many times when the free movement of content and dialogue impacts the corporate image (Youmans and York 2012). This, unfortunately, restricted the involvement of ordinary citizens and the potential contribution of grassroots communities to important and necessary conversations for disaster mitigation and recovery.

However, what the digital age has offered is the ability to place social networks online. No single movement relies on a dominant or individual member of a society to manage the knowledge and information disseminated across digital platforms. In fact, social media mirrors the offline networks. It requires trust and shared norms as a condition, and members of the network will then distribute information and knowledge, sharing the burden of doing so in advancement of the public good. It then transfers offline as those without access to broadband then disseminate knowledge online through offline apps like WhatsApp or through standard messaging systems that keep online networks connected to those offline (Earl et al. 2015; Wilson and Dunn 2011). It is a collective system, sustained by a collective effort that harnesses the ease of sharing knowledge, through the media platform. This development did not escape Haitian communities that have risen to the challenge of knowledge-sharing via digital media connectivity.

Still, it would be unwise to believe that such developments eliminate traditional forms of control. In reality, digital media connectivity exists within these forms of control. Social movements operate on platforms owned and controlled privately by multinational corporations that hold large amounts of economic influence and power. Social media platforms have the power to silence citizens through its algorithms and even silence presidents. Social media platforms are often unregulated or deliberately regulated, allowing for misinformation and disinformation to muddle the authentic flow of knowledge and to leave falsehoods unchallenged. Digital media also allow for propaganda and conspiracies to spread virally across the globe, weakening the media literacy abilities of citizens to discern truths from untruths (Benkler, Faris and Roberts 2018; Freelon and Wells 2020).

Digital media can therefore strengthen and bolster social capital, but it can also weaken trust within and across networks, while co-opting networks through disinformation. But its potential and demonstrated power to amplify the work of civil society still remains one of its unique aspects when harnessed in the right way. Its true potential is when it connects with the work of civil society offline, ensuring that digital advocacy is not simply “keyboard warriors” but results in real and tacit gains for the work of moving societies forward (Lopes 2014; Richardson 2020).

Democratizing Knowledge through the Mediated Public Sphere

When Haitians dismantled a three-decade-long dictatorship in 1986, it was the citizens who ushered in an era of democracy. However, that democratic exercise remains fragile and fledgling, and disasters are central to the instability of Haiti's democracy (Carlin, Love, and Zechmeister, 2014; Lin 2015; Pierre-Louis 2011) in specific ways. Disasters exacerbate the state's ability to advance an already unequal society, with large gaps between the rich and the poor (Mutter 2015). Developing nations must often look to foreign direct investment (FDI), either through state-to-state investment or private FDI to develop its sectors, train its workforce, and engage in the global marketplace. For developing countries that are vulnerable to natural disasters, FDI is then coupled with post-disaster management through the infusion of humanitarian aid, the global rebuilding sector through disaster-NGOs, and often multinational or bilateral arrangements around disaster relief and mitigation (Felbermayr and Gröschl, 2014). It prioritizes post-disaster aid in the larger framework of nation-building and empowers the donors of disaster aid in setting developmental milestones, often at the exclusion of local, non-political actors (Abrahams 2014; Concannon and Lindstrom 2011; Lin 2015).

The process of how post-disaster aid works and how it is controlled, earmarked, disbursed, and accounted for is often knowledge not readily and easily accessible to citizens (Tagliacozzo and Magni 2016). Often, citizens are unable to advocate for their needs or navigate resources because they lack access to information that allows them to determine what resources are available to them and how to retrieve them. Citizens often do not see disaster aid impact their daily lives. They learn of relief that the state receives but are not adequately empowered to account for that aid distribution in their communities and in their households. Local organizations play a key role in helping individuals and households navigate barriers to knowledge in accessing post-disaster resources (Nakagawa and Shaw 2004). They are often the go-betweens in accessing post-disaster aid from the state and distributing it to communities. When post-disaster aid distribution is effective, local organizations play a strong role in educating, advocating for, and empowering communities to receive resources to improve their short-term and long-term conditions after a disaster. When local organizations are either sidelined or marginalized as middlemen in aid distribution, post-disaster resources are often squandered, diverted, misused, or corrupted (Nakagawa and Shaw 2004).

Also, local organizations operate within the public sphere but require institutional bolstering to advocate on behalf of communities. Naturally, the state is the institution that recognizes local organizations and the role they play in serving communities at the grassroots level. However, when the state is corrupted, ineffective, or illegitimate, local organizations cannot rely on the state to play its role within a post-disaster society (Tierney and Oliver-Smith 2012). In the absence or compromised position of the state, other institutions become important conveyors of knowledge in post-disaster recovery (Tierney and Oliver-Smith 2012). The media assumes the role of a conveyor of knowledge within the public sphere, legitimizing sources or voices, facilitating debates, contextualizing data, seeking out truth, and holding the powerful accountable (Demiroz and Hu 2014; Sommerfeldt 2015). Local organizations and local non-state actors find redress in the media because media power bolsters their social capital, leveraging the trust, reach, and networks of community organizations in framing post-disaster discourse in ways that can be digested and activated by citizens (Sommerfeldt 2015).

Thus, the media support local organizations in their effort to filter the complexities of post-disaster aid in ways that allow citizens to engage in the public sphere about affairs that often impact and affect their daily lives but upon which they feel disenfranchised to act because knowledge is withheld, restricted, or confined to elites. Once the mass dissemination of knowledge through the media aids in citizen awareness and education, civic and grassroots organizations are able to restore power among citizens who can then rally around collective action because public consensus has reached a critical level (Brun 2018; Nakagawa and Shaw 2004). This educative role of the media in post-disaster contexts is crucial for citizen action and, in Haiti, can be seen in the ways citizens and civic groups achieve outcomes in holding other institutions of power accountable. Naturally, given the perceived slow recovery of Haiti, it would be helpful to see a more active public assume innovative and creative roles in empowering communities to action.

Haiti's Media and the Public Sphere in Post-Disaster Context

Over the decade since the 2010 earthquake, the role of Haiti's media models studies on the role of media in post-disaster contexts (Massey 1995; Perez-Lugo 2004). Before a disaster, media functions in the preparation mode, providing information necessary for citizens to make informed choices to protect themselves and their property from destruction. Certainly, disasters with shorter or no lead-time, such as earthquakes and tsunamis, restrict the capacity of the media to prepare citizens for disasters days or weeks prior. However, the media does play a "warning" role in providing ongoing citizen awareness of their environmental vulnerability and risk for disasters that can be sudden. During a disaster and in the immediate aftermath, the media functions in a social utility role. This is a linkage role providing citizens with comfort and solidarity that often allow individuals to overcome loss during a disaster and to develop resilience to face both the short- and long-term recovery ahead. After a disaster, the media enters a "response" role, providing information to access immediate resources to help citizens with the short-term needs of the post-disaster environment.

While scholars have often studied the role of the media within a few weeks to several months beyond a disaster, scholars in particular have yet to consider how many of these roles endure in protracted and prolonged post-disaster contexts. In many cases, the role of the media evolves in its preparation, social utility, and recovery functions because the effects of a disaster beyond a year become entrenched. In examining Haiti's media after the 2010 disaster, these traditional disaster media functions evolve to both advocate and hold accountable those in power, as the years and the decade after the earthquake unfold and as the state is unable to restore levels of relative stability that would allow for sustained recovery for citizens.

In the first half of the decade from mid-2010 to 2016, traditional media in Haiti, both commercial and alternative in particular, utilized both its social utility role and its response role to empower citizens with knowledge. However, it was a response that both advocated and sought accountability, and this is a key feature of Haiti's post-disaster media function that moves forward studies that theorize on protracted disasters. The fruits of those earlier mediated efforts can be found in the second half of the decade, where mediated spaces became digital public spheres that harnessed an informed citizenry in collective action aiming to advance the constitutional rights of Haitians to call for elections, electoral change, and governmental accountability.

Mediated Social Capital through Traditional Media in Post-disaster Haiti, 2010–2016

One of the key efforts of Haitian media in the first decade following the 2010 earthquake was the collective work of Haiti's Fourth Estate. Beginning in the hours after the disaster, sister radio stations relayed information to the provinces from the capital identifying meeting points for groups outside of the capital and affected areas to locate citizens in need of meals and to coordinate rescue efforts and to put up temporary dwellings.

Establishing Institutional Credibility and Trust Post-Disaster

Information sharing initially strengthened ties among different Haitian commercial media groups who signed a public declaration to work together in support of the public good given the magnitude of the disaster. The result of that collaboration was the Code of Ethics of the Haitian Press, signed on 8 December 2011 (UNESCO Office in Port-au-Prince 2011). The document was signed by the heads of the two leading commercial media organizations in Haiti at the time: the ANMH (L'Association Nationale des Médias Haïtiens) and the AMIH (L'Association des Médias D'Haïti Indépendants). Joining commercial media as signatories of the document were the leading community and alternative media organizations in Haiti at the time: SAKS (Sosyete Animasyon Kominikasyon Sosyal), REFRAKA (Rezo Fanm Radyo Kominotè Ayisyen), and MediaAlternatif. Additional signatories of the code were the Association of Haitian Journalists, SOS Journalists Association, and the Haitian Journalists Union Photographers. The code (2011) called for Haiti's Fourth Estate to hold the highest standards that would restore trust by citizens at a time when many institutions were under-resourced, marginalized, compromised, or corrupted. The code emphasized the freedom of the press, the verification and authentication of facts, the separation of facts from opinions, the respect of human dignity, vigilance with processing data and information, the correction of errors, the avoidance of rumors, independence from external influence or controls, electoral impartiality, and upholding brotherhood (UNESCO Office in Port-au-Prince 2011). The code therefore signaled to citizens and civic society that the media, in this post-disaster context, aimed to fill a void that was a missing institutional role, vital for aiding in the effective recovery of the country.

Accountability and Dissemination of Knowledge

Three of Haiti's alternative media organizations, SAKS, REFRAKA, and Alter Presse (the news agency arm of Media Alternatif, a media education/literacy organization), joined together with students and instructors in the Department of Communication at the State University of Haiti, Université d'État d'Haïti (UEH), to launch Ayiti Kale Je or Haiti Grassroots Watch. The investigative project produced thirty-nine dossiers that spanned from the latter half of the year of the earthquake until the work was concluded in 2014. The work of the alternative media collective was republished, repurposed, broadcast, and reprinted in the leading commercial media in Haiti, bolstering and amplifying the work of alternative and community media. The word cloud generated in Figure 4.2 is a sample of the featured dossiers from the end of the four-year project that reflects several key issues and the types of information the investigative media project targeted toward the education of citizens in the post-disaster years.

The project began its work by making its objectives and initial findings clear for citizens. It began its investigative work with the aim of answering

Figure 4.2. Ayiti Kale Je project. Word Cloud generated by Shearon Roberts using TagCloud application.

how Haiti will re-house 1.3 million people scattered across roughly 1,354 "squalid refugee camps." The project outlined its information seeking goal as follows:

> A dozen interviews, scores of documents, and many telephone calls later, Haiti Grassroots Watch discovered there actually does seem to be a plan. However, it is not readily accessible to the media or the Haitian public; it is so far only very loosely coordinated and, thus far, is not overseen by any Haitian agency or ministry, making accountability difficult, if not impossible. (Haiti Grassroots Watch 2010)

It then presents to audiences the following: "Who is in charge and what is the plan?" "What are the challenges?" and "Will it work." The report is disseminated in English, French, and Kreyol, and in text, audio, and visual format. The project took this simple approach to accountability journalism to seek answers to the most important post-disaster issues for citizens. Subsequent dossiers posed the same questions and provided answers for other post-disaster factors beyond homelessness covering cash allocations, elections, the cholera outbreak, restoring agriculture, unemployment, the reopening of business, stimulating exports, providing education access, deforestation, mining and the gold rush, hunger, and abortion.

Equally important to the topics, the project aimed to provide access to knowledge and information on who the project empowered to speak.

Ordinary citizens were elevated in every dossier to speak about their lived experiences, hardships, and fears in the years after the disaster. Additionally, Haitian organizations were given the platform to articulate the work they were doing and the ways they were marginalized. On the first anniversary of the disaster, Alterpresse reported that even international aid organizations had conceded that Haitians were "finding their own solutions" (Alterpresse 2011). International organizations had failed in their attempts to re-house over 800,000 displaced citizens, and it was Haitians themselves who were finding better solutions so they would not have to stay indefinitely in the squalid tent camps that had been set up by aid organizations. This was a remarkable source of pride for well-wishers of Haitian recovery. Haitians had worked to relocate to provinces outside of the capital, sourced money to repair homes to quasi-functional living status, combined living spaces with relatives and friends and/or simply moved their tents to safer, more communally coordinated neighborhood locations around the capital. The anniversary report also helped citizens contextualize why international aid does not always reach the source.

Additionally, the report noted that the international aid approach toward housing was ill-conceived because it was designed to help home-owners and failed to account for the fact that the majority of the displaced were renters. The report questioned international organizations that were leading the re-housing effort:

> The NGOs have \$100 million for "transitional shelters" . . . but the beneficiaries are without shelter. To whom should the NGOs be accountable, the donors or the would-be beneficiaries? What is or should be the role of the state, which seems more intent on focusing on its plan for 4,000 units in the Fort National neighborhood?
>
> And, after a "transitional" life in a "transitional" camp, will it be possible for the former renters—somewhere between 500,000 and 600,000 people—to save up one year's rent so they can move into something permanent? (Alterpresse 2011).

The project concluded its first anniversary analysis with the following critique that "Haiti Grassroots Watch considers the term 'non-governmental organization' or 'NGO' a bit of a misnomer because many 'NGOs' receive a great deal, and sometimes all, of their funding from governments" (Alterpresse 2011). In the case of this media project, this was truly the work of community stake-holders without considerable reliance on national or external organizational assistance.

The Haiti Grassroots Watch *Ayiti Kale Je* project distinguished international aid organizations as holding solutions that are not rooted in an understanding of social contexts prior to and during the disaster. It explained how Haitians are finding their own solutions to post-disaster challenges that the state is not willing or does not have the capacity to solve or that the international aid community fails to understand how to solve effectively. The project provided a space for citizens and civic networks to convey both their marginalization as well as their concerns that international aid solutions in post-disaster Haiti were neither native to the country nor sustainable, and in fact further formed dependencies that would be difficult to overcome once FDI was removed. This is precisely an observation that supports knowledge economies as a better model for development and recovery within the Black diaspora than other developmental models proposed by the West.

In Dossier 6, for instance, the Haiti Grassroots Watch (HGW) project examined the USAID program to provide Monsanto seeds to Haitian farmers, citing a concern about food insecurity and hunger post-disaster. The HGW project investigated the $20 million UN-led food program by interviewing farming groups. In its findings, the HGW team noted that some peasant farmer groups "receiving Monsanto and other hybrid maize and other cereal seeds have little understanding of the implications of getting 'hooked' on hybrid seeds." Other farmer groups did not know of the "health and environmental risks involved with the fungicide- and herbicide-coated hybrids" and "until Haiti Grassroots Watch intervened[,] they were planning to grind up the toxic seed to use as chicken feed" (Haiti Grassroots Watch 2011). The report quoted one farmer who noted that the seed had "little yield": "What I would like to tell the NGOs is that, just because we are the poorest country doesn't mean they should give us whatever, whenever," disgruntled Bainet farmer Jean Robert Cadichon told Haiti Grassroots Watch" (Haiti Grassroots Watch 2011).

The seeding report also helped citizens understand the lack of power and capacity the state held in intervening. The report noted that there were only two staffers that worked for the Ministry of Agriculture's National Seed Service. The report contextualized the challenges of the state by quoting the director of Haiti's National Seed Service who stated that since Haiti experienced severe disasters due to hurricanes the decade prior, the international NGOs and the FAO (the United Nations Food and Agriculture Organization) determined that they would exclusively fund and administer Haiti's seeding program dating back to 2008. Haiti's then director of the National Seed Service also pointed out the program could not be sustainable because the subsidies would only last for four years. The director, Emmanuel Prophete, told the HGW reporters that: "you have to ask yourself about the sustainability because if the policy changes one day, where will peasants get seeds? . . . We'll get to the point where, one day, we have a lot of seeds, and then suddenly, when all the NGOs are gone, we won't have any" (Haiti Grassroots Watch 2011).

For four years, the HGW project allowed citizens to have a deeper understanding of why billions of aid dollars did not reach or adequately serve them. The dossiers allowed Haitians to understand how post-disaster aid worked, who funded it, what the conditions for funding were, what was outlined and promised in donor plans, and whether international donor organizations met those objectives. The dossiers then assessed the work of international aid workers through bolstering citizens and civic groups on the ground to then evaluated whether the programs were working or not and how they could better suit the Haitian context. The HGW reports also helped Haitians to understand the constraints and limitations of their own government in executing and asserting control over disaster aid and provided space for government workers, where possible, to weigh in on the weaknesses of the state to intervene.

While the work of the dossiers was helped by independent, alternative media, the reports became the main news agenda items for mainstream commercial media. Popular radio talk programs in Haiti centered the reports and their findings as starting points for shows, interviews with elected officials, and Haitian experts who provided alternatives to the current approach to post-disaster recovery and development. Mainstream media served as a secondary site for alternative media in a rare show of solidarity lending its large, mass audiences to further distribute and contextualize the dossier's reports in the first half of the decade. It created a public sphere that was critical of the state of post-disaster aid delivery and laid the groundwork for the second half of the decade, which saw Haitian civic organizations pressure its government for accountability reports on the billions in international aid. The coordinating of Haitian society in civic protests was amplified through digital media but it was traditional and alternative/community media that served as a crucial, vital institution in the early post-disaster years, providing access to information and knowledge to aid citizens in seeking accountability and change in the years that would follow.

Mediated Social Capital through Digital Media in Post-Disaster Haiti, 2017–2021

The two hashtags #PetrocaribeChallenge and #Peyi-lòk, the first in 2018 and the second in 2019, defined the second half of the decade following the 2010 earthquake by demonstrating the protracted state of the failure of post-disaster recovery in Haiti. It also came after Hurricane Matthew in 2016 and other storms that reminded Haitians of their vulnerability and renewed the frustrations of how post-disaster aid does not mitigate the hardships disaster victims increasingly face. The hashtag #Petrocaribe-Challenge sought to use digital media to advance awareness from a simple question: Kot Kòb Petwo Karibe a? (Where is the PetroCaribe money?) The hashtag #Peyi-lòk took action on the ground. One digital movement aimed to raise awareness, the other pressured for accountability.

#PetrocaribeChallenge: Seeking Awareness

"Knowledge is power" is a well-used saying that applies to the #PetrocaribeChallenge. The movement that started digitally and then resulted in massive protests over four months, coincided with a government report on what happened to roughly $2 billion from the Venezuelan initiative in place between 2008 and 2016. Haitians had received access to information through the "Rapport de la Commission Sénatoriale Spéciale d'Enquête sur le fonds Petro Caribe Couvrant les Périodes Annuelles Allant de Septembre 2008 A September 2016" (Report of the Special Senatorial Inquiry Commission on the PetroCaribe funds for the annual periods spanning September 2008 to September 2016) (2017).

The report noted that three governments failed to properly manage the funds, awarded contracts to companies that did not adhere to ethical bidding processes, and misrepresented accounting of the exchange rates and the funds, some of which may have been stolen and were still unaccounted for. Although the report was released in 2017, Haitians soon began to experience the crunch of inflation and rising fuel prices in 2018 that brought the issue into full focus as it added even more hardships for citizens. Haitians took to the streets calling for President Jovenel Moïse and implicated government officials to be prosecuted (Nugent 2019). The delay in when the report was released and when Haitians began protesting the following year allows for a better understanding of how digital media played a role in globalizing Haitian social networks, specifically in knowledge awareness, more so than knowledge acquisition. Certainly, knowledge acquisition had taken place the year prior with the release of the Senate report, a result of the work of civic organizations and media in the years before aiming to seek answers not only about PetroCaribe dollars but the roughly $14 billion in aid from the international community since the earthquake. While Venezuela's FDI was state-to-state and the UN's Haiti Reconstruction Fund was multilateral, Haitians and Haitian organizations had been shut out and marginalized from the effective administration of aid from both approaches.

Haitian society had already been asking questions about international aid prior to the protests of 2018 and the use of the hashtag, which was first issued by Haitian-Canadian activist Gilbert Mirambeau, Jr. However, questions about Petro Caribe dollars and Haiti Reconstruction dollars existed, long before the hashtag trended, among local organizations, citizens, and traditional media. Therefore, the #PetroCaribeChallenge movement was not based in knowledge access and acquisition. That existed prior to the 2018 protests, marked specifically by the 2017 Senate special report. The #PetroCaribeChallenge was aimed at building external solidarity for Haitians.

Social capital is transformative when it is bolstered and allied with other powerful institutions. Global media was co-opted to bring international pressure and attention to the Haitian state for squandering funds it could control, compared to the UN-administered Haiti Reconstruction Fund, which often flowed through US aid programs and NGOs. PetroCaribe dollars, on the other hand, were managed directly through the Haitian state, and as such, Haitians could hold their government accountable for the mismanagement of this state-to-state FDI. Since Haitians are acutely aware of the power that foreign states play in bolstering Haitian governments, the protests were aimed at discrediting the current Haitian government among global citizens, making it unpopular for citizens to support their government's position to accept the status quo of leadership in Haiti.

An analysis of the first 200 random Twitter results for the #petroCaribeChallenge, filtered from 1 January 2018 to 1 January 2019, indicated that the hashtag was primarily used by accounts that posted with English text. When using the accompanying hashtag #KotkobPetwoKaribeA, the first 200 tweets primarily used French text, followed by English text, with Haitian Kreyol as a distant third.

Marc-Henry Pierre and Mehmet Güzel's 2021 study of the #PetroCaribeChallenge as political communication also supports this finding. The researchers examined a sample of 163 purposely sampled tweets using the hashtag #KotKobPetwoKaribeA and found that the majority of the posts between December 2019 and January 2020 were in French, and the majority were merely retweets, carrying little information or details about the movement and the information it aimed to provide or featured only a small number of comments engaging in a digital dialogue around the hashtag. The most important discourse of the hashtag was to accuse the Haitian government of corruption, the researchers found. In applying media political economy and political communication theory, Pierre and Güzel argue that "the critical analysis of the Twitter-based Petro Caribe Challenge revealed a movement that excluded the majority of the Haitian population as most of the selected tweets are written in French . . . . Despite the political content of the tweets, the lack of comments and reactions to the tweets hinders communication between protesters" (Pierre and Güzel 2021).

While Pierre and Güzel see little evidence of engagement from Haitians in the country itself with the digital aspects of the movement, local media coverage showed that the hashtag was prevalently showcased in demonstrations and protests. Certainly, Haitians may not have widely utilized social media platforms to post the hashtag, but the hashtag's relevance and symbolism in the movement was prevalent offline as well. This is because digital movements do not require every citizen to participate but representative actors, who then translate digital activism through offline social networks. This does not delegitimize the digital nature of movements; it only speaks to its limitations when large mass publics are unable, due to digital access constraints, to fully engage a movement online.

More important to the discussion of whether the #PetroCaribeChallenge was widely used in Kreyol online is the concern for who its intended audience was. Haitians already had knowledge of the misappropriation of PetroCaribe dollars before 2018 and certainly prior to the 2017 Senate report, which already confirmed what Haitians perceived was mismanagement of the Venezuelan fund. The hashtag was not a message among Haitians; it was a message from Haitians to the outside world, signaling that Haitians intended to seek a change in governance and wanted to amplify this messaging given the foreign influence exerted in Haitian elections, particularly since the earthquake. Haitians themselves did not need to engage in this digital messaging. They only required the Haitian diaspora, its "10th department," and transnational allies to amplify the messaging. However, on the ground, protests prepared the country for what was to come, which was the collective shutting down of the country in 2019 in order to pressure the government to step down and hold elections. Therefore the #PetroCaribeChallenge movement did not seek to acquire knowledge, it aimed to share awareness over digital platforms and gain solidarity for collective action, which was regime change in the country.

#Peyi-lòk: Seeking Change

While the #PetroCaribeChallenge was largely marked by digital media use in French and English, #Peyi-lòk (Country shutdown) was declared in Haitian Kreyol and used as such in 2019. Haitians forced the closing of schools, government offices, transportation, and much more, bringing the country's economy to a grinding halt for ten days in February 2019. The lockdowns would continue throughout 2019 as Haitians protested energy shortages as a result of the end of the discounted fuel from Venezuela given that country's own economic challenges. Peyi-lòk intensified the civil disobedience actions of the #PetroCaribeChallenge in defiance of a government that refused to step down, hold itself accountable, and retaliated against protesters resulting in the death of dozens of citizens who participated in the movements. The outcomes of Peyi-lòk, compared to #PetroCaribeChallenge, were specifically geared to internal outcomes. It prompted the senators and representatives of the government opposition to call for the president to step down for the "stabilité et la protection des vies et des biens" (for the stability and protection of lives and well-being) (Pierre 2019). The parliament also demanded a change in power that followed the constitution and put in place a transition.

Chambers of Commerce then supported the call by citizens for change and called for a *dialogue national* (national dialogue) among all diverse sectors, which would be a conversation among Haitian sectors and built on trust and transparency. The religious community, led by the Catholic church, also spoke for the need for political change and reform, citing the growing misery of citizens. Peyi-lòk demonstrated the importance of a collaborative exercise, maintaining that in using digital and traditional media, other powerful institutions in Haiti would come on board as a check and balance on political institutions. The movement saw economic elites, the religious community, and the opposition join forces with the media and civic organizations to demand change for the sake of the country.

Social networks in Haiti were truly empowered when multiple institutions reached critical consensus about the state of affairs in the country and what was needed to change the status quo. It first started with educating the public sphere at the beginning of Haiti's recovery after the disaster, and the role of the media was central in the acquisition, contextualizing, and dissemination of knowledge that allowed civic society to sustain a decade-long effort to bring about social and political change. This effort continued even after the assassination of President Moïse in 2021 and the continued inability of the interim state to hold free and fair elections. However, Haitian citizens are fully aware of the power of their social capital and continue to leverage it as multiple sectors across the country continued convening in 2022 to keep the pressure on the state to fully follow its constitutional duty to hold elections and to begin the difficult work of building a more equitable and sustainable Haiti for all its citizens.

Conclusion: Non-Western Knowledge Economies

For a country that is often labeled as impoverished, Haiti is rich in social capital. What this chapter outlined is that when social capital is empowered through knowledge access, knowledge acquisition, and knowledge dissemination, citizens are equipped to collectively bargain and lobby their state in advancement of both the macro and micro well-being of the nation. Social capital cannot operate in isolation either. It needs institutional validation and recognition in pursuit of national transformation. When the state or other powerful forces marginalizes civic groups, other powerful social institutions are needed to ensure that social networks can build national consensus, coordinate national action, and sustain the kinds of pressure that would result in change, particularly for economies that have suffered under decades of systemic strain, as is the case in Haiti. In other words, social capital requires empowerment in order to achieve other forms of capital gains.

The media remains a social institution, that when societies require it, can work to aid the public sphere in building consensus. The information gathering and disseminating role of the media makes it powerful as an agent in building knowledge acquisition in societies often shut out from decision-making and disenfranchised by a fledgling democratic state. While the media may not often work in the best interests of citizens at all times, prolonged and protracted disasters force media systems to aspire to a greater public good, because its very nature as a social institution rests on its value to be of service and need to citizens as both a source and a platform.

This is even more relevant in our digital media landscape where citizens can often bypass traditional media systems and access information directly and use media networks to disseminate information and contextualize it without the filter or use of traditional media. And while new media platforms can often usurp the ability of traditional media to set the agenda and to frame discourse, disasters prove that traditional media will still have a role to play. This is because, often, the platforms that are still able to operate in a disaster are often analog and offline, like terrestrial radio or in some cases standard messaging systems. Not every citizen is able to engage digital platforms equally, even as broadband penetration grows globally or in developing nations where disposable income does not allow for vast use of the internet or mobile applications. Citizens will still require traditional media, which has a far wider penetration, to convey and relay digital conversations in a more widely accessed public sphere. What digital discourse offers, particularly in the Haiti case, is a global awareness of why *èd pa mache* (aid does not work), sensitizing citizens abroad to understand how the actions of foreign governments and international organizations further aggravate the ability of a country to self-determine and seek its own sustainable path without perpetuating forms of economic dependency that have kept Haiti impoverished in the modern era.

More importantly, institutions are able to hold other institutions accountable. Media discourse can educate and activate citizens who then hold officials accountable as they take to the streets in protest, or when they vote for electoral change. It introduces into the public sphere ways of knowing and builds understanding around a path forward that is rooted in native ways of knowing that allow citizens to navigate the opportunities and risks of the environments they inhabit. Without institutional power, the social capital of a citizenry remains a latent force. It becomes transformative when it can leverage the networks within it to advocate for systemic change and that often first starts with an educated, aware, and empowered citizenry.

Shearon Roberts holds a PhD in Latin American and Caribbean Studies from Tulane University's Roger Thayer Stone Center in Latin American Studies. She has been a fellow with Duke University and University of North Carolina Chapel Hill Consortium in Latin American and Caribbean Studies and a faculty research fellow with the University of Southern California's Annenberg School's Center for Public Diplomacy where she publishes on global Black racial social justice movements. She has conducted extensive field research in Haiti, Cuba, the Dominican Republic and Trinidad and Tobago. She is a former journalist.

References

Abrahams, Daniel. 2014. "The Barriers to Environmental Sustainability in Post-disaster Settings: A Case Study of Transitional Shelter Implementation in Haiti." *Disasters* 38(1): S25–S49.

Alterpresse. 2011. "What Is the Re-housing Plan?" *Haiti Grassroots Watch*, 12 January. http://haitigrassrootswatch.squarespace.com/5logementEN.

Benkler, Yochai, Robert Faris, and Hal Roberts. 2018. *Network Propaganda: Manipulation, Disinformation, and Radicalization in American Politics*. Oxford: Oxford University Press.

Bojarski, Sam, Vania André, and Garry Pierre-Pierre. 2020. "Social Media and Its Role in Connecting Earthquake Survivors within the Diaspora," *Pulitzer Center*, 10 January, Retrieved 24 April 2023 from https://pulitzercenter.org/stories/social-media-and-its-role-connecting-earthquake-survivors-diaspora.

Brun, Ashly. 2018. "Haiti: The Relationship between Political Instability and Post-disaster Response." Master's thesis. Worcester, MA: Clark University.

Carlin, Ryan E., Gregory J. Love, and Elizabeth J. Zechmeister. 2014. "Natural Disaster and Democratic Legitimacy: The Public Opinion Consequences of Chile's 2010 Earthquake and Tsunami." *Political Research Quarterly* 67(1): 3–15.

Chafee, Zechariah. 1947. *A Free and Responsible Press: A General Report on Mass Communication: Newspapers, Radio, Motion Pictures, Magazines, and Books*. Chicago: University of Chicago Press.

Concannon, Jr., Brian, and Beatrice Lindstrom. 2011. "Cheaper, Better, Longer-Lasting: A Rights-Based Approach to Disaster Response in Haiti." *Emory International Law Review* 25(3): 1145–91.

Demiroz, Fatih, and Qian Hu. 2014. "The Role of Nonprofits and Civil Society in Post-disaster Recovery and Development." In *Disaster and Development: Examining Global Issues and Cases*, ed. Naim Kapucu and Kuotsai Tom Liou, 317–30. Cham: Springer.

Earl, Jennifer, Jayson Hunt, R. Kelly Garrett, and Aysenur Dal. 2015. "New Technologies and Social Movements." In *The Oxford Handbook of Social Movements*, ed. Donatella della Porta and Mario Diani, 355–66. Oxford: Oxford University Press, 2015.

Felbermayr, Gabriel, and Jasmin Gröschl. 2014. "Naturally Negative: The Growth Effects of Natural Disasters." *Journal of Development Economics* 11(1): 92–106.

Freelon, Deen, and Chris Wells. 2020. "Disinformation as Political Communication." *Political Communication* 37(2): 145–56.

Haiti Grassroots Watch. 2010. "What Is the Plan for Haiti's 1.3 Million Homeless?" *Haiti Grassroots Watch, Dossier 1, Haiti's Homeless*, 12 October. Retrieved 24 April 2023 from http://haitigrassrootswatch.squarespace.com/haiti-grassroots-watch-engli/2010/9/27/what-is-the-plan-for-haitis-13-million-homeless.html.

Haiti Grassroots Watch. 2011. "Seeding Reconstruction or Deconstruction?" *Haiti Grassroots Watch, Dossier 6, Seed Distribution*, 30 March. Retrieved 24 April 2023 from http://haitigrassrootswatch.squarespace.com/haiti-grassroots-watch-engli/2011/3/30/seeding-reconstruction-or-destruction.html.

Hudson, Dale, and Patricia Zimmermann. 2015. *Thinking through Digital Media: Transnational Environments and Locative Places*. New York: Springer.

Hughes, Sallie, and Chappell Lawson. 2005. "The Barriers to Media Opening in Latin America." *Political Communication* 22(1): 9–25.

Institute for Justice and Democracy in Haiti (IJDH). 2018. "On #humanrightsday, @BAIhaiti held discussions housing as a human right and how support for the #KotKobPetwoKaribeA movement turned into a state massacre. Read more here on how grassroots solidarity is defending human rights in Haiti." *Twitter*, 10 December. Retrieved 24 April 2023 from https://twitter.com/ijdh/status/1072264849692991488?s=20&t=u2EhZLp8yD98k0IeunnpEA.

Jenkins, Henry, and David Thorburn. 2003. "Introduction: The Digital Revolution, the Informed Citizen, and the Culture of Democracy." In *Democracy and New Media*, ed. Henry Jenkins and David Thorburn, 1–17. Cambridge, MA: MIT Press.

Jorba, Laia, and Bruce Bimber. 2012. "The Impact of Digital Media on Citizenship from a Global Perspective." *Digital Media and Political Engagement Worldwide: A Comparative Study*, ed. Eva Anduiza, Michael J. Jensen, and Laia Jorba, 16–38. New York: Cambridge University Press.

Juris, Jeffrey S. 2005. "The New Digital Media and Activist Networking within Anti-corporate Globalization Movements." *The ANNALS of the American Academy of Political and Social Science* 597(1): 189–208.

Lawson, Chappell. 2002. *Building the Fourth Estate: Democratization and the Rise of a Free Press in Mexico*. Berkeley: University of California Press.

"Le rapport PetroCaribe de la Commission Sénatoriale Spéciale d'Enquête du Sénateur Evallière Beauplan" [Report of the Special Senatorial Inquiry Commission on the PetroCaribe funds for the annual periods spanning September 2008 to September 2016]. *Haïti Liberté*, 11 November 2017. Retrieved 24 April 2023 from https://haitiliberte.com/le-rapport-petrocaribe-de-la-commission-senatoriale-speciale-denquete-du-senateur-evalliere-beauplan/.

Lesser, Eric L. 2000. "Leveraging Social Capital in Organizations." In *Knowledge and Social Capital: Foundations and Applications*, ed. Eric Lesser, 3–17. Woburn, MA: Butterworth-Heinemann.

Lin, Thung-Hong. 2015. "Governing Natural Disasters: State Capacity, Democracy, and Human Vulnerability." *Social Forces* 93(3): 1267–300.

Lopes, Amandha Rohr. 2014. "The Impact of Social Media on Social Movements: The New Opportunity and Mobilizing Structure." *Retrieved 24 April 2023 from* https://www.creighton.edu/fileadmin/user/CCAS/departments/PoliticalScience/Journal_of_Political_Research__JPR_/2014_JSP_papers/Lopes_JPR.pdf.

Lugo, Jairo. 2008. *The Media in Latin America.* London: McGraw-Hill Education.

Massey, Kimberly B. 1995. "Analyzing the Uses and Gratifications Concept of Audience Activity with a Qualitative Approach: Media Encounters during the 1989 Loma Prieta Earthquake Disaster." *Journal of Broadcasting & Electronic Media* 39(3): 328–49

Mirambeau, Jr., Gilbert. 2018. "Kot kòb petwo karibe a???" *Twitter,* 18 August. Retrieved 24 April 2023 from https://twitter.com/GibszZZz/status/1029445853788991490?s=20&t=SopH-GTvrBoZP5rgXRTE3w.

Montas-Dominique, Michele. 2002. "The Role of the Press in Helping Create the Conditions of Democracy to Develop in Haiti." *University of Miami Law Review* 56(2): 397–403.

Mutter, John C. 2015. *The Disaster Profiteers: How Natural Disasters Make the Rich Richer and the Poor Even Poorer.* New York: St. Martin's Press.

Nakagawa, Yuko, and Rajib Shaw. 2004. "Social Capital: A Missing Link to Disaster Recovery." *International Journal of Mass Emergencies and Disasters* 22(1): 5–34.

Nugent, Ciara. 2019. "Why a Venezuelan Oil Program Is Fueling Massive Street Protests in Haiti." *Time,* 24 June. Retrieved 24 April 2023 from https://time.com/5609054/haiti-protests-petrocaribe/.

Perez-Lugo, Marla. 2004. "Media Uses in Disaster Situations: A New Focus on the Impact Phase." *Sociological Inquiry* 74(2): 210–25.

Pierre, Gotson. 2019."Perspective: Haiti: pays-lock, acte VI," *Alterpresse,* 12 February. Retrieved 24 April 2023 from https://www.alterpresse.org/spip.php?article24038.

Pierre, Marc-Henry, and Mehmet Güzel. 2021. "A Critical Analysis of the Public Sphere in the Digital Age: The Case of the Petrocaribe Challenge Movement." *Etkileşim* 8 (2021): 90–107. https://doi.org/10.32739/etkilesim.2021.4.8.139.

Pierre-Louis, Francois. 2011. "Earthquakes, Nongovernmental Organizations, and Governance in Haiti." *Journal of Black Studies* 42(2): 186–202.

Putnam, Robert D. 1993. "The Prosperous Community: Social Capital and Public Life." *The American Prospect* 13: 35–42.

Richardson, Allissa V. 2020. *Bearing Witness while Black: African Americans, Smartphones, and the New Protest #Journalism.* New York: Oxford University Press.

Roberts, Shearon. 2016. "Then and Now: Haitian Journalism as Resistance to US Occupation and US-Led Reconstruction." *Journal of Haitian Studies* 21(2): 241–68.

Schuller, Mark, and Pablo Morales, eds. 2012. *Tectonic Shifts: Haiti since the Earthquake,* vol. 43. Sterling, VA: Kumarian Press.

Sommerfeldt, Erich J. 2015. "Disasters and Information Source Repertoires: Information Seeking and Information Sufficiency in Post-earthquake Haiti." *Journal of Applied Communication Research* 43(1): 1–22.

Stieber, Chelsea. 2020. *Haiti's Paper War: Post-independence Writing, Civil War, and the Making of the Republic, 1804–1954.* New York: New York University Press.

Tagliacozzo, Serena, and Michele Magni. 2016. "Communicating with Communities (CwC) during Post-disaster Reconstruction: An Initial Analysis." *Natural Hazards* 84(3): 2225–42.

Tierney, Kathleen, and Anthony Oliver-Smith. 2012. "Social Dimensions of Disaster Recovery." *International Journal of Mass Emergencies & Disasters* 30(2): 123–46.

UNESCO Office in Port-au-Prince. 2011. *Kòd deyontoloji medya ak journalis peyi d Ayiti* [Code of ethics for media and journalism in Haiti]. Retrieved 24 April 2023 from https://unesdoc.unesco.org/ark:/48223/pf0000371669.

Wagner, Laura. 2017. "Nou toujou la! The Digital (After-)Life of Radio Haïti-Inter." *Archipelagos* 2. https://doi.org/10.7916/D8J394ZN.

Wilson, Christopher, and Alexandra Dunn. 2011. "The Arab Spring: Digital Media in the Egyptian Revolution; Descriptive Analysis from the Tahrir Data Set." *International Journal of Communication* 5: 1248–72.

Youmans, William Lafi, and Jillian C. York. 2012. "Social Media and the Activist Toolkit: User Agreements, Corporate Interests, and the Information Infrastructure of Modern Social Movements." *Journal of Communication* 62(2): 315–29.

CHAPTER 5

Rwanda

Sociocultural Practices for Resilience and Recovery

SAMUEL HABIMANA AND EMMANUEL BIRACYAZA

The incidence of wars and other disastrous stressors has long had negative impacts on human lives. Disaster leaves behind unmentionable grief to all involved and usually diminishes the quality of life of an entire population across all ages and generations as they are forced to endure in the aftermath of the disaster. So how do societies cope? How do they institute programs aimed at recovery and reconciliation and build resilience among communities as they search for development and sustainability? In this examination of catastrophes in context, the case of Rwanda offers ideal scrutiny into recovery practices that have stood the test of time.

With a population of only 13 million, Rwanda is the smallest eastern African country. Rwandans also share one indigenous language, Kinyarwanda, which everyone uses. Despite this, Rwanda experienced a fierce civil war, a disaster of enormous proportion, in the 1994 genocide against the Tutsi. The consequences of this disaster loom to this day. The mental health of genocide survivors, their descendants, and perpetrators of the genocide, along with economic deprivation, social and community distrust, and physiological problems have challenged the political, economic, and social development of this territory for the last twenty-eight years. However, in the last five years or so, World Bank reports (2023) indicate that growth within the last decade in Rwanda has averaged 7.2 percent while per capita Gross Domestic Product (GDP) grew at 5 percent annually. The 2020 recession induced by the worldwide COVID pandemic (in which GDP fell by 3.4 percent) was the country's first since 1994. Such developmental performance, albeit described in purely economic terms, conveys the hope to onlookers that the well-being of Rwandans in the aftermath of their political, cultural, and social crisis contributes to viability and sustainability for progress in the country. To combat the effects of the genocide in the immediate aftermath of the disaster, however, Rwandans made a significant effort to respond and manage the pressures experienced in their lives. Several homegrown solutions or adopted interventions had to be used to restore resilience and wellness within the Rwandan community. Community-based interventions and person-based therapies were extensively implemented to address the impacts of those stressors.

This chapter strives to explain Rwanda's post-genocide recovery strategies. Given the diversity of human responses to stressful life events, particularly traumatic events, which we ordinarily expect to overwhelm people's coping resources, the government of Rwanda created several initiatives to promote unity and reconciliation, effective reintegration of genocide prisoners and ex-prisoners, reduce intergenerational legacies of genocide, and empower the local leaders, teachers, and church leaders so that they could contribute to the reduction of the genocide effects in Rwandans of all generations. These efforts, made in partnership with other health (national and international) organizations, demonstrate the potential for increasing resilience, social identity, reconciliation, social healing, post-traumatic growth, reintegration of genocide perpetrators, and reduction of trauma transmission from one generation to another in post-disaster Rwanda.

To heal the nation and return to a semblance of harmony among formerly hostile groups, integrative psychotherapeutic approaches have been adopted to reflect Rwandan culture, assert respect for personal loss within communities and dispose of victims in a traditional, dignified, and respectful manner. Rwanda's achievements to date have been a lighthouse for managing trauma and restoring justice in the post-disaster period in remarkable ways. In the following sections of this chapter, we will discuss reasons for the genocide, justice restoration strategies, community-based mental health interventions, social arts and culture inputs, government programs and policies, and religious and traditional rituals.

Rwanda's Catastrophe

Circa 1990 and later, Rwanda experienced wars from insurgencies in Northeastern Rwanda. There is no agreed single cause of the invasion and political disturbance factors that led to the 1994 genocide against the Tutsi. Political instability and bad governance were, perhaps, foremost among suggested causes of the genocide that ensued in Rwanda. Some scholars have attributed Rwanda's catastrophe, however, to its colonial past, arguing that decades of division and incitement to hostility against the Tutsi by radicals in the Rwandan leadership structure (see Heldring

2021) controlled by members of the Hutu majority, contributed to the breakdown of peace and harmony among the two groups. Few deny that an intentional process of smearing the Tutsi minority as inferior and dangerous, contributed to the perception among Hutus that the only solution to the danger Tutsis posed was to eradicate their threat to the community.

In approximately one hundred days from the first open murmurings against the Tutsis, government forces, militias, and ordinary citizens took up arms against the Tutsi and killed more than one million of them, in addition to perpetrating sexual violence against the women and girls of that social and ethnic group. Journalists writing for human rights organizations and the press were quick to highlight Belgian colonial policy, which favored the Tutsis as superior to the Hutus and better able to manage positions of leadership. Thus, after more than thirty years of currying favor, there was little appetite among Hutus to stop the anti-Tutsi hatred that had developed.

In the chronology of events that led to the genocide of 1994, Rwanda experienced a coup d'état in 1973 by a Hutu general that reinforced anti-Tutsi and pro-Hutu sentiments, leading to another wave of violence against the Tutsis for no other reason than to make clear that the latter were in no way superior to the Hutus. In hindsight, therefore, it was not at all surprising that the Tutsi, many of whom had been exiled to surrounding areas of Rwanda, pushed back against the violence and demanded the right to return to their place of birth. Persistent claims of rights echoed cries for democratic leadership in Rwanda and brought the crisis to a head with all-out genocide against the Tutsis opening a new chapter in the history of Rwanda. Lessons, subsequently learned, have led to eagerness over the years to erase the memory of the past by transitioning to a shared future where group identities are absent from educational, government, and social institutions of Rwanda. No longer do school children have to reveal their group affiliations, their identification documents (e.g. passports), nor do they have to label passport-bearers as one ethnic group or another with an identification brand.

Indeed, after the genocide, the Rwandan community experienced the impacts of the catastrophe in a variety of ways, including cultural, environmental, economic, social, psychological, and physical. The effects of the genocide manifested on the individual level, community level, and societal level. The Rwanda National Commission for the fight against genocide reported that in Rwanda after only three months of clashing, more than a million Tutsi ethnic minorities were killed by the extremist Hutu ethnic majority. The rate of post-traumatic stress disorder (PTSD) was found to range from 50 to 76 percent among women, children, and men exposed to genocide (Pham, Weinstein, and Longman 2004). Furthermore, the economy was devasted by the wanton destruction of infrastructure. Rwanda also experienced the human impact of loss from a high incidence of prisoners, a massive number of orphans and an abundance of single parents. All those co-social demographics indicated that not only the economy but also the Rwandan way of life needed to undergo a full, societal overhaul.

Post-disaster recovery is best served through building the resilience of the people, community by community. Homegrown strategies, developed in every sphere of life (cultural, social, economic, political, etc.) have been among the most innovative methods for seeking sustainability. These strategies are mostly rooted in the knowledge capital resident in communities, where the narratives and memories of the past offer lessons on best practices that may still yield hope for a sustainable future. In this regard, this chapter will unpack some of the mostly homegrown, rather than imported, restoration practices, which include strategies for realizing a just transition to peace and harmony in post-disaster Rwanda, innovative socio-therapy practices for mental well-being, creative public policy for strategic leadership and reconstruction, sociocultural practices for generating reconciliation, apology, forgiveness, renewal, and so on.

Restorative Justice

Community Justice

In the aftermath of a national catastrophe, such as the genocide in Rwanda of 1994, a just transition to normalcy requires innovative strategies for realigning communities. Home-based practices in judicious settlement emanating from the knowledge capital found in community membership offer a realistic expectation that harmony may be restored among survivors and perpetrators when communities use traditional methods to ensure free and open communication across all parties. In addition to the contributions of health care providers who treated those wounded in the genocide for the many mental disorders observed as outcomes of the genocide, Gacaca courts and Gacaca judges were selected from the communities to restore resilience and bring reconciliation to genocide survivors and perpetrators. Gacaca is a system of community justice, which is rooted in Rwandan knowledge traditions. It is patterned on the idea of truth and reconciliation, arguing that when communities face their problems as a community, punishments are likely to match the weight of the crimes as recognized by the communities in which both perpetrators and victims reside. Gacaca courts constitute the traditional judicial meetings of the Rwandan communities where, prior to the genocide, people gathered to discuss property and family problems. At those gatherings, traditional leaders had the last word, judicially. By applying traditional mechanisms (such as Gacaca courts) to mete out justice to demoralized communities, Rwandans hoped to effectively complement conventional judicial systems with homegrown methods, with a real potential for promoting justice, reconciliation, and a culture of democracy. In 1998, this traditional model—used to roll out a transitional type of jurisdiction to end the long history of unpunished violence, try the mass crimes in a community-based way, tackle the humanitarian issues in prison, find the truth about what had happened, determine the level of punishments, and bring about reconciliation and unity—was seen as the most innovative method for restoring justice in war-ravaged Rwanda. These are in harmony with other indigenous African knowledge systems adopted by African countries (Krog 2015a) and could well serve the needs of communities within the African diaspora experiencing and managing civil unrest in post-disaster periods.

Given the enormity of the civil catastrophe in Rwanda, transitional jurisdiction was restricted to cover justice in the immediate aftermath of the 1994 genocide. Crimes to be tried before the renewed Gacaca courts had to be focused on uncovering the truth of what took place during the genocide and determining the appropriate level of punishment. Significant crimes related to a history of unpunished violence committed by those who masterminded the genocide, were referred for trial to the International Criminal Tribunal for Rwanda in Arusha, Tanzania. Crimes committed by the coordinating perpetrators were referred to the national courts. Those accused of rape faced the highest class of prosecution. Overall, in Rwanda, the Gacaca court has been acclaimed for contributing much to the psychosocial healing, reconciliation, compensating of genocide survivors for their property loss during the genocide, honorable burial of the bodies of the genocide victims, repair of properties damaged during the genocide, and increasing the level of self-reliance among all people. These strategies are similar to the mechanisms implemented in South Africa (Koulen 2009; Krog 2015b) in their post-apartheid era and in Zimbabwe following independence from colonial management and could continue to be applied in other areas of the diaspora after similar civil catastrophes.

Contextually, the strategy for Rwandan justice, manifested in homegrown practices such as Gacaca, where the hearing in local trials of persons violating community and cultural norms, was brought before a joint committee, justifies the value of relying on the historical tradition, whereby facing one's accusers allows for the management of conflict in a way that the community finds helpful. Additionally, education and training in academic and social institutions have aided such methodologies in repairing the social identity hindered by the genocide and conflicts preceding the traumatic events of 1994. Recent generations specializing in anthropology, psychology, public health, social work, and more encourage policy decisions centering on crimes against women, for example, and their issues in a sociocultural environment where inequities are addressed and solutions found internally. As anthropologists interpret traditional practices, psychologists also weigh in with knowledge-based strategies to promote mental health through innovative formulas, relieving stress, and reducing collective and transgenerational trauma. These new methods supplement conventional practices that have also contributed to mental health. Practices such as laughter yoga, traditional dance, and handcraft have had a substantial impact on the mental well-being of communities. These practices have been artfully designed to work at the community level to engage necessary conversations and behaviors that improve peaceful communication and *Ubuntu* (radical hospitality) within neighborhoods.

Mental Health Interventions

Community Resilience Model (CRM)

The African diaspora is renowned for its resiliency. Given the history of slavery and cultural disruption in the lives of African-descended peoples, community resilience has been the cornerstone of sustainability for the diaspora. In Rwanda, CRM interventions are psycho-biologically based healing approaches applied in communities with the aim of significantly improving mental health and achieving community well-being (Freeman et al. 2021; Grabbe et al. 2020). In Rwanda, as in other sub-Saharan African countries to have experienced traumatic events that hampered the lives of past and current populations, this intervention was implemented and successfully boosted the resilience of individuals and promoted psychosocial healing through the training in the CRM skills.

Application of such skills helped to create "trauma-informed" and "resiliency-informed" individuals and empower communities that share a common understanding of the impact of trauma and chronic stress on the nervous system. Knowledge of said skills/approaches inform how resiliency can be restored or increased when a skills-based model is applied.

Resilience is the ability to bounce back from some real, experienced, difficulties or the skill of utilizing knowledge that can restore or establish inner strengths and resources for overcoming seriously adverse, traumatic events and their impacts, thus allowing individuals to pursue and succeed in their endeavors (Ionescu 2011; Julien-Gauthier and Jourdan-Ionescu, 2015). Resilient individuals, when experiencing well-being, might become better than ever in embracing positive well-being. The 1994 genocide sur-

Figure 5.1. Socio-therapeutic training exercise. Source: RRGO project © Samuel Habimana

vivors showed resilience and continued living and achieving their goals after the genocide. An example of this is explained in a study conducted on students associated with Genocide Survivors Students Association (Association des Etudiants Et Éleves Rescapés Du Genocide; AERG). These students were seen to show resilience and were reported to have attained their vision of achieving goals for living successfully in the present and even in the future (Dushimirimana, Sezibera, and Auerbach 2014). Resilience is often measured by the extent to which accomplishments surpass original intentions and lead to the adoption of a new lifestyle. Studies show that survivors of the Holocaust achieved resiliency by thriving in the present and remain mentally stable despite the rate of trauma they experienced in their post-disaster era. The influence of resilience varies across different sectors of one's life, including social, individual, political, economic, and cultural interventions (Luthar and Cicchetti 2000).

Other scholars have observed how traumatized communities demonstrate resiliency. Resilient individuals maintain a physical equilibrium of living in the present and achieving a stable mental balance in the face of difficulty (Miller-Karas 2015, 2018). Resilient individuals tend to develop new possibilities and positive thinking, which indicate post-traumatic growth (Anderson et al. 2019; Lepore and Revenson 2006; Ogińska-Bulik and Kobylarczyk 2016). Based on the post-disaster experiences of Rwandan community members, where life was inevitably stressful, and there was tremendous variability in how people responded to and managed life's stressors, it can be argued that the introduction of coping skills served to boost the resiliency levels of communities. Knowledge sharing has the capacity to avail communities of the potential for resilience. Diasporic people may be empowered to develop and adopt skills of resilience and gain post-traumatic growth that has been important to their everyday memories and plans for the future. In a recent study, a relationship between resilience and PTSD showed stability was robustly and significantly associated with a decreased likelihood of mental health problems (Bogopolskaya 2019; Wrenn et al. 2011). Strength is exemplified by refreshment, mental equilibrium, and homeostasis of the nervous system (Miller-Karas 2018). Across Africa and its diaspora, there has been a notable dissociation in periods of civil unrest between post-traumatic stress disorders and resilience in individuals and communities.

The interventions of five CRM skills, namely tracking, resourcing, grounding, gesture, and spontaneous movements, now help and demonstrate that this biologically based model significantly contributed to the restoration of mental equilibrium and mental well-being in genocide survivors who had previously displayed trauma symptoms. The genocide survivors increased their ability to show love, embrace a relationship, and heal from their trauma after participating in the Community Resilience Model skills training program (Habimana et al. 2021). Further, Rwandans, within their own understandings, have applauded the resilience of the Rwandan community after the genocide. It is still difficult to understand and differentiate the definition of resilience and factors of resilience among different communities. The experiences encountered by people originating in Africa highlight the risk factors of disequilibrium of the nervous systems, as perceived in victims, survivors, and descendants.

The patterns of this dysregulation of the nervous system increase the risk of physical and psychological illnesses such as immune-system disorders, depression, anxiety, and cognitive impairment. For example, some survivors may have the chronic impact of trauma. Some communities may experience chronic collective trauma. If those communities of people are not provided health interventions such as the CRM skills that may target the regulation of the nervous system, the life of the affected may be worsened (Pfefferbaum et al. 2013; Miller-Karas 2015). Based on principles of participatory action research, CRM applications had the potential to contribute to community resilience by encouraging and supporting community participation, reconciliation between survivors and perpetrators, and build cooperation, communication, self-awareness, humanity, and trust among Rwandans. Therefore, somatic interventions that specifically target the way post-traumatic responses are stored or patterned in the body, in addition to working with cognitions and emotions (Ellis and Dietz 2017) are useful strategies in the search for sustainable development in post-disaster communities.

Clinically, we can say that psychotherapeutic approaches in the aftermath of disaster effectively restore the nervous system, which is the major mechanism that allows other health systems to work well. Based on Rwanda's experiences, communities in the diaspora may advocate for CRM skills training after confronting a physical (weather-related) disaster, a sociopolitical catastrophe, or any other form of traumatic event that may impede their development and sustainability as a nation. Clearly CRM can aid self-care for those community members who are frontline workers, first-responders to crises, or live in highly traumatized and marginalized communities. Such recommendations concur with studies that found CRM skills training to be a critical approach to knowledge production that addresses psychopathologies and helps communities to be resilient and work in equilibrium (Leitch and Miller-Karas 2009).

Socio-therapeutic Practices and Mental Well-Being

Prior to 2005, several interventions to resolve post-disaster underdevelopment were focused on the individual. Several Rwandans, for example, received patient-centered interventions and hospital-based therapies that aided in biopsychosocial healing and, as a result, were indirectly supported in reconciliation, social cohesion, and reintegration. However, due to a cultural taboo, many Rwandan communities at first opposed mental health interventions.

Not dissimilar in this resistance is the way African diasporic communities also believe that mental health treatment is anathema to their well-being and is likely to further contribute to the marginalization rather than recovery of individuals of African descent undergoing such treatment. The exigencies of disaster, however, call for desperate efforts to be implemented to ensure recovery and sustainable development. Thus, socio-therapy was introduced into Rwanda as a community-based approach to promoting a high level of reconciliation between genocide survivors and perpetrators, as well as formulating effective strategies for reintegration of genocide prisoners and ex-prisoners, reduction of the intergenerational legacies of genocide, and empowering the local leaders (see Figure 5.2 below) to take action to reduce the effects of the genocide. Note that socio-therapy strategies differ from psychotherapy because socio-therapy incorporates the totality of the clinical setting. The listed principles and methods are

Figure 5.2. Gathering of genocide survivors, perpetrators, and family members. Unity and Reconciliation program.

applied as a model of confronting patients with their "outside world." This is the opposite of facing a patient with an inside world during psychotherapy (Jansen et al. 2015; Richters, Dekker, and Scholte, 2008). Hence, this practice is implemented via community-based interventions, in various countries, to address the impact of conflicts, war, and genocide. Notably, African countries have benefited from such interventions, which have resulted in historical healing from trauma and promoting individual and community resilience in the post-conflict period. Such interventions have been implemented in African countries, such as Uganda, South Africa, and Sudan (Brewer 2020; Krog 2015a), but have not been necessary in post-disaster conditions in the Black diaspora. Given the recent post-election experiences in regions of Guyana following months of debate about who had won the elections, socio-therapy may well become necessary in future to defuse the rising passions that stem from the ethnic undertones of community conflict in the city and rural areas.

In Rwanda, community-based socio-therapy has also been used as a hybrid intervention integrating psychological trauma healing with rebuilding community trust and resilience. Within a group of survivors, participants were able to transition from one exercise in recovery to another. Distinct sequential stages of the healing process in socio-therapy include "safety," "trust," "care," "respect," "new life orientation," and "memory." The approach has often promoted psychological and community resilience through shared storytelling. Storytelling has had many benefits. Sharing their stories allows participants to reflect on the actions of the past and determine the origins of the behavior that led to the hostility resulting in hatred for each other and genocide in retaliation for mutual anger. Knowledge sharing also allows participants to determine better ways to recover, move forward, and grow to achieve the best outcomes for the whole nation. As argued by George Santayana (1905), and later supported by Winston Churchill in a 1948 speech to the British House of Commons that those who fail to learn from history are doomed to repeat it. Rwandan communities that revere the lessons of the past have gained much in terms of individual and national growth by not repeating the mistakes of their recent past. Communities in the Americas and the Caribbean (especially Guyana in recent years) should be similarly encouraged not to forget the consequences of political disasters of the past, lest they repeat them.

In the aftermath of political conflict and following transitions to peace and democracy, Rwanda has attempted to rebuild political structures and shape social relationships to create more peaceful futures. Yet, the memory of past violence does not stay neatly in the past. It festers and disrupts and haunts our best efforts to move forward. Along with significant social and political actions, it is crucial that peacebuilding processes also grapple with and respond to the individual and collective memory ghosts of past violence (Bayer 2010; Dudouet 2007).

The approach of socio-therapy as a clinical effort was adapted to the Rwandan context for dealing with the effects of the genocide among the genocide survivors and their descendants, and genocide prisoners and ex-prisoners and their descendants. The approach aimed at promoting an effective reintegration, reduction of transgenerational legacies of genocide, and empowerment of local leaders, teachers, and church leaders to find solutions to problems associated with the effects of genocide (Biracyaza and Habimana 2020). Often, families entered the therapeutic setting, in socio-therapy, with complex trauma histories and intergenerational trauma that may persist for a long time when not addressed using the appropriate interventions such as community-parent-based and community-youth-based interventions. These histories are complicated in that every generation seems to be affected by adversity, mental health issues, substance use, and government agency involvement. In agreement with scholars researching this topic, it was found that untreated traumatic stress has severe consequences for children, youth, adults, and families. These are congruent with other studies documenting that public health interventions are substantial for empowering people and promoting their well-being (Goodman 2013).

Other studies have also made strides in understanding the impact colonialism, slavery, war, and genocide have on families and what can be done to address this in the context of a present problem (Brewer 2020). To be clear, not all families carry the burden of what happened generations ago; some families are more resilient than others and can acknowledge the past while looking toward the future. This chapter on post-disaster recovery and development aims to address the varied impacts of disaster while providing ways to work with families who present with complex histories of disaster in their own lives. It is helpful to take this into account, since it is the hope that communities in the African diaspora, with their diverse experiences of disaster and resultant trauma may apply the best practices that may alleviate their situation in their own regional spaces. While socio-therapeutic practice may, overall, serve as a helpful interventional approach, because of its contribution to social cohesion within the same neighborhood or group, it must be recognized that variation in its impact may result. In Rwanda, a group of ten to fifteen participants sitting in a circle to talk about their experienced health stressors and the effects on their psychosocial health (Richters et al. 2008) has been impactful for the community in which the strategy was applied. The intervention helped to restore the psychosocial well-being of the participants and community and contributed to economic welfare. Participants also talked about events related to their traumatic experience, be it sexual or gender-based violence, war, genocide, or other public health concerns, in addition to the destruction of trust and safety within their communities, as being akin to "life without humanity." To facilitate a sense of redress for people in Rwanda, an approach was needed to address psychological factors operating at individual and community levels. Analyzing the PTSD associated with the 1994 Rwandan genocide against the Tutsi as one of the root causes of domestic violence in Rwanda, with some research postulating that domestic violence has been a vector for the transmission of trauma to children in the aftermath of the genocide, may not be consistent with approaches and findings for Louisiana following the devastation of Hurricane Katrina in 2005. This illustrates the fact that recovering from disaster, given the variety of its forms, does not and cannot have a one-size-fits-all treatment plan.

Still, sharing socio-therapeutic knowledge and practices with other communities experiencing trauma related to idiosyncratic disaster experiences is particularly helpful to recovery and a better mental and social functioning approach to the problems at hand. The benefits of information sharing, as it relates to disaster recovery, emanate from the shared dialogues (Biracyaza and Habimana 2020) within communities and have a powerful impact on the continuing search for recovery and developmental sustainability.

Religious and Cultural Practices

Religiosity in Rwanda

Before the genocide, traditional healers, pastors, and herbal medicine were used to treat mental health problems. The study, conducted in the Northern Province of Rwanda, reported that traditional healers use a variety of indigenous labels to describe what biomedical psychiatry categorizes as psychotic disorders (Tan et al. 2021). These are associated with various explanatory models, from supernatural and/or spiritual causes to somatic causes such as HIV (Patel 2011). Research on traditional healers and mental health revealed that 9 percent of Rwandan respondents consulted traditional healers, and 11 percent consulted a religious or spiritual advisor in the aftermath of the genocide. In the sample analyzed, it was found that consultation with conventional healers was more likely to be predicted by (older) age, (black) race, unemployment, (lower) education, and anxiety or substance use disorder (Sorsdahl et al. 2009).

Cultural Practices

Of course, such findings demonstrate the range of cultural practices engaged in the aftermath of post-traumatic stress events. Disasters, as well as any other disruption to communities in the diaspora, are often met by a return to roots, to the conventional wisdom, to seek understanding, new and old ways of knowing, so that action can be taken to mitigate the risks to recovery, development, and a better future. Attentive community leaders are aware that clinical assistance is only one way to heal communities traumatized by disaster and do their best to engage government action to provide other forms of sustenance for marginalized members of the community, especially those who are unable to afford medical care or may be distrustful of modern and technological means of treating trauma. Uniting behind cultural norms helps to aid healing and recovery in fundamental ways. This is a lesson that diasporic communities in the Global South understand well. Diasporic communities in developed localities of the Global North, who have long forgotten or abandoned the cultural rituals of their ancestors, find slower healing for the lack of homegrown practices to which they can resort. The latter thus continue to depend on government assistance, costly therapy, available charitable services, or, in the absence of those, give in to the wanderings of their mind and seek shelter and food wherever they may find it, sometimes resorting to alcohol or drug abuse to "kill the pain" of survival.

Government Policies

Reconciliation Strategies

Reconciliation after a disaster such as the genocide in Rwanda is a priority in nations struggling to recover and grow. In addition to the cultural and medical strategies made available to disrupted communities, governments have stepped up to mitigate the fallout from the disaster. In the case of Rwanda, attention was given, over the decades since the genocide, to provide diverse community-based health interventions in the countryside. The aim of these efforts was to promote the well-being of Rwandans and increase the level of social identity. Although these interventions have knowingly contributed to psychosocial healing, social cohesion, trauma healing (individual and collective trauma), they have also contributed to reconciliation and forgiveness journeys among the genocide survivors and perpetrators. As shown in prior studies (International Peace Institute 2013), forgiveness and reconciliation are the essential processes leading to individual and community resilience. Reconciliation and forgiveness are the significant steps in peacemaking, collective healing, and social cohesion. Reconciliation has been the final step of forgiveness since they are complementary. Therefore, the government of Rwanda has established a month of unity and reconciliation that aims to promote recovery, effective reintegration of genocide prisoners and ex-prisoners, and promote the well-being of all Rwandans, including genocide survivors and perpetrators. This policy is put into effect on 31 October every year.

Several other policies are implemented to foster community and family resilience. Among these are such activities as constructing a house for vulnerable people, empowering youth and reintegrating them in activities that promote reconciliation, campaigning in the community to encourage resilience in Rwandans, eradicating the genocide ideologies in youth and adults, involving Rwandans in the effects of discrimination and genocide by promoting the Ndi Umunyarwanda program, translated as "I am Rwandan," and performing community service to promote a unique identity and social cohesion. This line of policymaking was introduced on 1 October 2020, when Rwanda launched a month-long event commemorating the Unity and Reconciliation of Rwandans on the thirtieth anniversary of their journey to restoring Rwanda and liberating itself from the shackles of discrimination and division.

Since 2007, a particular time has been set aside to reflect on unity and reconciliation every year. Initially, a week was assigned for such reflection but in 2018, as Rwanda approached the twenty-fifth year since the genocide, the commemoration was increased to a month-long program

Figure 5.3. Government provided green space available for Umuganda activities. © Samuel Habimana.

at the request of Rwandans. Each year has a theme. For instance, in 2020, the theme of this remembrance was "Let's work together to build a country without discrimination." The people of Rwanda recognize all these sociocultural events or memorials as a new, collective step in the country's recovery and the achievement of self-determination and sustainable development. Rwanda fully recognizes that reconciliation is a genuine pathway to healing and with the recovery of individuals comes a national product that can be sustainable. The government of Rwanda in listening to the people and seeking to meet their demands has added currency to global speculation that Rwanda is leading the way toward overcoming disaster and attaining a sustainable future, free from inequity and unrest.

One often overlooked fact is that the government of Rwanda looks nothing today as it did thirty years ago when government leadership and opinion practitioners comprised mostly of Hutu males who stirred up "identity conflict" and added vengeance to their policymaking arsenals. In the post-genocide period, with a society where more women than men were available to take the reins of government, a new vision was reflected in Rwandan decision-making and women were empowered to nurture communities back to good health. With careful planning and collaborative use of the knowledge capital produced in towns and villages, reconciliation and other healing strategies were implemented. Representation of women in government, as discussed more fully in other parts of this text, has been a major source of progress and development in this region of Africa and a light for those in the diaspora (as seen in Barbados in 2021) who are fast catching up with Rwanda's lead in female representation in the hallways of government.

Umuganda

Umuganda, a program for restoration and patriotism, is a pre-colonial Kinyarwanda word that means working together to support and reconstruct impoverished conditions. According to Penine Uwimbabazi (2012), in pre-colonial Rwanda, Muganda was a traditional practice and cultural value of working together to solve social and economic problems for mutual benefit. This practice aimed to support the people with low income and in need of immediate financial support (Uwimbabazi 2012). The activities of Umuganda included, for instance, cultivating the fields for those who were unable to do so due to either physical handicap or old age, building houses and spaces for their livestock, carrying the food, and fetching the water for physically disabled people and providing transportation to medical facilities for those who were in need. Today, green spaces, as illustrated above, are available for Umuganda activities, which may be conducted at the individual or household level. These played a significant role in protecting human security and increasing household income. This community practice started in pre-colonial times, and it was characterized by bringing together community members to solve social and economic problems for mutual benefit (RGB 2017).

After the genocide, communities, with the support of government officials, returned to the practice of Umuganda, which has the potential to restore resilience and reconciliation among survivors and perpetrators of the genocide. The village codes to build the house of survivors were re-established and the perpetrators' families join others to give support. In some families, the perpetrators do more than others to demonstrate their regret and a willingness to foster cohesion in the community. Furthermore, survivors show willingness to apply this program to support the families of perpetrators. This mutual collaboration significantly contributes to the community's restoration of wellness. Umuganda activities have a significant impact on conviviality within Rwandan society. As a cultural-based approach, Rwandans embrace the opportunity to contribute to the health of the Rwandan community in support of economic advancement and communal cohesion, which fosters both a spirit of reconciliation and patriotism.

Reconciliation Villages

For those who survived the Rwanda genocide or watched as their loved ones were slaughtered or raped, reconciling with the genocide perpetrators once seemed unimaginable. Living together in harmony and helping each other was also unbelievable. To promote neighborhood safety and reconciliation between the genocide survivors and perpetrators, reconciliation villages were constructed so that genocide survivors and perpetrators could live side by side. Such villages were built by the Rwandan government with the support of the Prison Fellowship Rwanda (PFR) organization. This organization, PFR, has adopted a variety of approaches to implement healing and reconciliation programs, including programs titled Peacebuilding and Reconciliation, Psychosocial Healing and Reconciliation, Human Rights Promotion and Legal Aid, and Spiritual Resilience (Mafeza 2013). The reconstituted Rwandan government fully supported the communities in efforts to make these programs successful. Most of the labor of love, however, came from community advocacy, primarily the diligence of women, some of whom worked in tandem with local female political representatives, who served as government liaisons to the community. Under the Peacebuilding and Reconciliation program, the PFR (2003) started the reconciliation villages program. Compared to other existing villages, these reconciliation villages provided an enabling environment for continued practical reconciliation while also offering shelter to the target groups and an opportunity to bring peace and reconciliation to the community.

Sociopolitical Perspectives

In postcolonial Africa, several countries from this continent built on universal norms and enshrined them in domestic legislation and practices as they attempted to reconcile the imperatives of national independence with adherence to international law. Some countries experienced violence, wars, and genocide that decimated the quality of life of individuals and communities. The continent also witnessed the prevalence of undemocratic and dictatorial regimes that were characterized by gross violations of human rights, extrajudicial executions, and violent change of power. Although some of the countries that experienced those traumatic events have been struggling with the psychiatric and social effects of those events, Africa has promoted reconciliation, social cohesion, social identity, constitutionalism, trauma healing, resilience, self-reliance, and traditional justice in several countries (Bruch et al. 2009; International Peace Institute 2013). Africa achieved this through diverse efforts to incorporate value and norms into national, regional, and continental structures and strategies to attenuate the harmful effects of those events.

Rwanda has not been the only African territory to benefit from these cultural efforts to rebuild its society after a disaster. African countries, such as Uganda and Burundi, faced civil conflicts that brought several negative impacts to the lives of the people and social cohesion among the community members. After these widely publicized harmful events of the 1980s, Rwanda's experience with the 1994 genocide that killed millions of people and brought extreme poverty, mistrust, family conflict, orphans, widows, psychopathologies, and collective trauma added new fears that development in African regions was being retarded rather than sustained. Across the Black diaspora as well as in other parts of the world, many gave in to the negative perspective that Africa was doomed and its people, no matter where scattered, lacked the capacity to self-govern efficiently. Few considered the global challenges facing Africa and its descendants, the discrimination and inequity that brought countries like Rwanda to this state of psychological collapse. Few envisaged a return to a commitment to responsible governance with fairness, equity, and respect for the sanctity of human life. But, a few years later, Africa momentously contributed to reducing the negative impacts of lifelong trauma by elaborating and articulating the norms that have contributed to the well-being of the people and development. Although the conflicts experienced were due to colonialism, weak governance was also one of the factors that led to the battles, war, and genocide. So different organizations and government leadership collaborated to bring peace and restore stability to African countries. The contributions of Africa, by creating initiatives favorable to peacebuilding (McNamee and Muyangwa 2021), may serve as a beacon for other areas of the African diaspora with whom such understandings can be shared. Recognizing that development in all its forms can be sustained in the Black diaspora is a lesson in resilience this volume hopes to reiterate.

In addition, government policies aimed at building the resilience of Rwandan communities and reinforcing the wisdom of pursuing self-determination rather than continued political and economic dependence on a colonial past. One of the policies instituted in Rwanda, Ndi Umunyarwanda, translated as "I am Rwandan," is specifically relevant to ideals of self-determination. It is a political move aimed at nationwide "de-ethnicization" of the population and it has taken many forms. In recent times, this program identified Rwandans as unique and renounced the colonial strategy of divide and conquer. Policy leaders made the argument that there is no ethnic identity of groups under labels such as Hutu, Tutsi, or Twa. Thus, the Ndi Umunyarwanda program, established in the post-

Figure 5.4. Ndi Umunyarwanda Gathering. © Samuel Habimana.

genocidal period to strengthen reconciliation, patriotism, and national identity, has engendered Rwandans' humanity and social identity. The program also initiated the building of a national identity based on trust and dignity. Its aim was to strengthen unity and reconciliation among Rwandans by providing a forum for people to talk about the causes and consequences of the genocide and what it means to be Rwandan.

Consistent with other government initiatives, officially aimed at overcoming tribalism and building a national identity, Ndi Umunyarwanda is also about sharing a single narrative of the country's past, present, and future (Kagoyire and Richters, 2018; Richters and Kagoyire, 2014). Ndi Umunyarwanda is the policy established in 2013 after the community-based Gacaca courts were concluded. The government of Rwanda based this strategy on restoring the unity of Rwandans and promoting community resilience. No conventional means existed to hold accountable ordinary Rwandans who committed genocide. There were too many cases; nearly two million Rwandans were implicated. So, many programs were applied to create unity, resilience, and reconciliation among Rwandans, post-genocide (see Figure 5.4 above).

Based on how the genocide was conducted and its impacts, it was not easy to imagine that people could live together, work together, study, and pray together. If as many victims as perpetrators conflict, justice becomes difficult. For example, South Africa experienced social injustice during the period of apartheid. Due to the traumatic past experienced in South Africa, South Africans thought that forgiveness and reconciliation were impossible. This also occurred in Rwanda, where genocide survivors and perpetrators did not realize that reconciliation could be possible among the genocide survivors, their families, genocide perpetrators, and their families. But in several instances, the unforgivable has been forgiven by envisioning a new national identity (Gallagher 2002).

Moreover, current strategies also contribute to effective reintegration of genocide prisoners and reduction of their sentences. So, the Gacaca judges began the process of healing and forgiveness by facilitating a resolution of the problems related to the genocide. Via community services made available to Rwandans, not as punitive judgments but as restorative practices, what was feared to be impossible, became possible. To better appreciate the particularities of restorative justice in the Rwandan context, it is necessary to understand the rationale behind it and examine its implementation. Many Rwandans now believe that being Rwandan within the community empowers the society to work for the betterment of the whole nation rather than allow for the recovery of individuals. This perspective sustains the belief that when knowledge capital is tapped into and produced at the community level, the empowered society can effectively advocate for its whole well-being and benefit from sustained development.

Gender Equality

It cannot be denied that bottom-up strategies are more likely to succeed in collaboration with government support. At the heart of Rwandan community success in achieving peace, harmony, and economic security in an environmentally just society, are the rights, roles, and duties assigned to women in the post-genocide era. As stated earlier in this text, the war in Rwanda left many more women than men alive to take the reins of government and restore peace and progress to Rwanda. According to the Rwanda National Institute of Statistics, Rwandan women represent 58.6 percent of the population and more than 63 percent of the members of parliament in Rwanda are women. Those statistics illustrate how gender equality and equal participation in policy making in Rwanda contributes to positive reconfigurations of Rwanda after the long-term effects of disaster.

After the genocide, with more women surviving the atrocities and heading households in Rwanda, the promotion of gender equality was the right choice for helping the nation develop and recover from the effects of the 1994 genocide. Such developments are not sufficiently known in the

African diaspora. There continues to be a pressing need for knowledge sharing among a people who are constantly reminded by the West of their perceived failure to live up to European standards of civilization. It cannot be overstated that when women engage in policy making, they are less likely to ignite wars than to generate policies that nurture and rebuild communities.

Social Arts and Culture (SAC)

Sociocultural influences in Rwanda have included sports, music, laughter, exercise, and dance throughout the ages. In the aftermath of the genocide perpetrated there, the creative arts have played a meaningful role in recovery. In Rwanda, traditional practices that potentially promote mental health have been identified and explored. For example, there are ways to sing songs that heal the wounded heart and restore mental wellness. These are songs typically sung when a person has died or when a person's health is impaired and in need of special care. Communities in Rwanda have not only revived these for small family gatherings but have embellished and shared them with larger communities and international visitors. When performed at memorial ceremonies and other national events, these social arts convey, with a level of certainty, that a national spirit is alive and thriving in Rwanda. Although such behavior is not unique to Rwanda, it gives hope to many who wrote the country off as non-viable in the post-disaster era. As observed in airports and other public spaces in the Caribbean and Africa, diasporic countries exuding a warm welcome to visitors do so with song and dance and genuine cheer.

Sports Activities

Additionally, sporting exhibitions were revived by the government of Rwanda to promote the well-being of Rwandans. It is considered an essential method for building resilience and bringing positive feelings and emotions to communities of people. After a tragedy, such as the genocide of 1994, these recovery methods were used among families of survivors to bring hope for children and older adults who survived the atrocities of the day. In concurrence with previous studies that documented the effectiveness of sport and songs on mental health outcomes, this approach to recovery aided a reduction of depression, anxiety, and trauma. It also increased the level of social cohesion (Büssing et al. 2012; Pascoe et al. 2020). It is evident that these practices helped rebuild the social life of Rwandans and heal wounds from the genocide.

Contact sports, such as football and volleyball, are worthy of mention here. After the genocide, it was hard to motivate people to meet again with feelings of goodwill. Sporting opportunities were devised to bring back happiness and well-being to the Rwandan community. Sporting events were held to reconnect the Rwandan community and eliminate ethnic segregation. Sports were encouraged in Rwanda in all population categories, including youth. Rwanda began identifying champions among primary, secondary, and tertiary institutions. Schools met and played different games: football, volleyball, basketball, and handball. Adults also participated in tournaments and won at the intersectoral or district level. Those games signaled the togetherness and wellness of the Rwandan community in the post-genocide period.

Engaging in sports was also another way to teach and share social values, including fair play, respect for opponents, and the importance of hard work. It also taught the significance of teamwork, cooperation, and inclusion. Thus, through sporting activities, sponsors of recovery reinforced the benefit of helping participants acquire values and life skills consistent with positive social relationships, collaborative action, and mutual support. All those collaborations have contributed to making Rwanda live again. For example, when the sports organization known as Rayon Sports F.C. plays and wins at national and international football games (now ranking fourth in Rwanda NFL), a high percentage of the Rwandan community displays togetherness and joy in the shared experience.

Dance

Traditional weddings and dancing also played a special role in the post-disaster recovery of Rwanda. The Rwandan wedding remains a symbol of unification between families. It is also a way of sharing daily and of protecting family relationships. Marriage ceremonies have served as an excellent sociocultural practice that brings communities together. Music and dance are part of the Kinyarwanda tradition; special dances are performed for leisure, entertainment, and celebrations like marriage. The Kinyarwanda dancing (*Intore*) is unique; it consists of heavy drums carried and played by energetic men and women who dance. Sometimes, spectators sing along so that the performance becomes colorful, interactive. The Kinyarwanda dance demonstrates their tradition as the dance motions illustrate their way of life, such as cattle rearing and personal bravery. Performances educate younger audiences and international visitors on the traditions and way of life of Rwandans.

Dancing and weddings in Rwanda also contributed to the restoration relationship within the Rwandan communities destroyed because of the genocide. When a son from the perpetrators' family marries a female survivor, they invite both families to attend the wedding. These cultural experiences have made it possible for Rwanda to overcome the conflicts resulting in the genocide and attest that acknowledging the past helps one never to want to repeat the atrocities of the past.

Laughter Yoga

Laughter yoga has been used in several countries. In 2012, a team from Australia brought laughter yoga to Rwanda. This intervention has successfully motivated the Rwandan community to strive for wellness and well-being after exposure to the adversity of genocide. Through laughter yoga, Rwandans bounced back and shared smiles, joy, and peace. Laughter is contagious! It does not discriminate among individuals; rather, it generated wellness and well-being throughout the Rwandan community. In exercises in laughter yoga, conducted for the benefit of international visitors and also practiced in community groups and on campus, it is illuminating to see the powerful impact laughter has on the individual, the community, and others affected by the contagion of laughter and the energy it fosters. It is a satisfying exercise for people of all ages. In societies where there is strife, poverty, and many other vulnerabilities, sharing the gift of laughter is precious indeed.

Rites and Respect

After the genocide against the Tutsi, genocide survivors and their descendants developed heart wounds, complex trauma, and some experienced psychosomatic symptoms. Not finding the bodies of loved ones killed in the genocide was devastating for surviving members of society. This loss maimed their personal growth, disrupted collective healing, and stymied sociocultural development. Despite the fact that many (discussed above) strategies were implemented to contribute to reconciliation and psychosocial healing, the problem of not burying the bodies of genocide victims became a public health concern that hampers the well-being of survivors and perpetrators (African Rights and REDRESS 2008; Korman 2016). It is disturbing to see that many years after the genocide, the remains of genocide victims are still being retrieved and given to families for a decent burial.

In general, when a person dies in Rwanda, the rest of the family has to bury them to honor the bodies of the genocide victim. Most of the time, after losing the victim, the other family member(s) may develop psychiatric ailments that may last a long time. Some psychiatric disturbances may include depression, anxiety, and inconsolable grief. This perspective is expressed in recent scholarship, establishing that death is a "universal, natural, persistent, inescapable, unavoidable, and undeniable fact of life" (Ekore and Lanre-Abass 2016). When a death occurs, there is usually a strong negative impact on the family and friends of the deceased. The magnitude of the impact often depends on whether the end was expected or unexpected. Such effects may be economic, emotional, psychological, or socioeconomic, depending on the relationships between the dead and those left behind. In the case of genocide, however, the circumstances of death are elevated. The number of lives lost constituted a disaster of enormous proportions and impacted families, the nation, and the Rwandan diaspora in the wake of so many lost lives, heightened economic insecurity, the collapse of families, and devastation in the physical environment. Although some scholars highlighted the impact, the deaths caused entire Rwandan communities to be crippled by fear, anger, and grief, which effectively impacted the whole society.

According to Copernicus, the human soul perishes with the body at death, bringing all sensation and conscious existence to an end (Ekore and Lanre-Abass 2016). Although some scholars or philosophers believe that death is natural, and there is no need for it to have such an impact, the relatives who were closer to the dead suffer deeply from the senselessness of the genocide and fear, anger, and grief, once the inevitability of death becomes apparent and intolerable. In the case of Rwanda, the loss of so many lives and souls brought an abrupt end to the potential for personal, economic, and environmental security and halted national development.

These experiences of Rwanda also support the view expressed in previous studies that sharing collectively is better than sharing individually because collective memory has the potential impact of reconciliation and social cohesion (Rafferty 2017). The fact, though, is that collective memory is a massive barrier to conflict resolution in societies that endure intractable conflicts. It is for this reason that sharing the experiences of this disaster and the solutions found to mitigate the risks of such a disaster being repeated is recounted here. Members of the African diaspora, wherever currently located, can learn from the innovative measures applied in Rwanda to create their own narratives of identity, responsible self-governance, and the successful pursuit of sustainable development for their communities.

Conclusion: Balancing Resilience with Development

The genocide against the Tutsi in Rwanda utterly destroyed the standard of living for Rwandans, and several repercussions still act as obstacles to social reconstruction and human welfare. To overcome such effects, many sociocultural practices and innovative policies were established in an attempt to mitigate the impact of that disaster, ensuring that Rwanda achieved the viability that all nations require in order to grow and prosper. Overcoming negatives in a country devastated by colonial exploitation, injustice, racial degradation, ethnic discrimination, natural resource rape, a high rate of poverty and economic insecurity, environmental distress, physical and psychological illness, and mistrust of policy and opinion leadership throughout Rwanda, is an impossible task for any country to undertake. Diverse creative policies and psychological models adapted from homegrown and foreign anthropological and culture-based practices were set in place to build the resilience of Rwandan society and recover from the harmful effects of genocide. Rwanda serves as a crucial example to other countries, demonstrating the truth that it is possible to live again after a calamity as horrific as genocide by employing tried-and-true methods to discover solutions. However, resilience has not yet reached its peak; much is still left to be done for development to be achieved and for Rwandans to exercise self-determination in the hope of becoming a sustainable and prosperous force for development.

Samuel Habimana is the co-founder and Executive Director of Rwanda Resilience and Grounding Organization. Currently, he is a PhD student in Social Welfare and Social Research at the School of Behavioral Health, Loma Linda University. Habimana earned a Master's degree in Public Health and a Bachelor's degree in Clinical Psychology from the University of Rwanda. He is a teacher of the Community Resiliency Model (CRM) and a Laughter Yoga Ambassador. Mental health, resiliency peace building, green space, and community dynamics are among his research interests. Habimana is a researcher and international consultant on building resilience and social cohesion.

Emmanuel Biracyaza is a PhD student at the Université de Montréal under supervision of Prof. Martine Lévesque and Carolina Bottari. His research aims to generate in-depth knowledge about the realities, constraints, and strengths affecting parenting, in a Rwandan perspective, multilevel interventions, and actions that may support parenting and strengthen family capacity. Emmanuel graduated from the University of Rwanda with Bachelor's and Master's degrees in Clinical Psychology and Public Health, respectively. Prior to pursuing doctoral studies, he coordinated the socio-therapy program at Prison Fellowship Rwanda. Behavior change, occupational therapy, parenting, child development, psychosocial well-being, and conflict resolution are among his research interests.

References

African Rights and REDRESS. 2008. *Survivors and Post-Genocide Justice in Rwanda: Their Experiences, Perspectives and Hopes.* Retrieved 24 April 2023 from https://redress.org/wp-content/uploads/2018/01/Nov-08-Survivors-and-Post-Genoicde-Justice-in-Rwanda.pdf.

Anderson, Kimberley, Amra Delić, Ivan Komproe, Esmina Avdibegović, Elisa Van Ee, and Heide Glaesmer. 2019. "Predictors of Posttraumatic Growth among Conflict-Related Sexual Violence Survivors from Bosnia and Herzegovina." *Conflict and Health* 13(1): 1–11. https://doi.org/10.1186/s13031-019-0201-5.

Bayer, Reşat. 2010. "Peaceful Transitions and Democracy." *Journal of Peace Research* 47(5): 535–46. http://www.jstor.org/stable/20798924.

Biracyaza, Emmanuel, and Samuel Habimana. 2020. "Contribution of Community-Based Sociotherapy Interventions for the Psychological Well-Being of Rwandan Youths Born to Genocide Perpetrators and Survivors: Analysis of the Stories Telling of a Sociotherapy Approach." *BMC Psychology* 9: 1–15. https://doi.org/10.1186/s40359-020-00471-9.

Bogopolskaya, Natalya. 2019. "Relationship between Resilience Factors and Presence of Posttraumatic Stress Disorder Symptoms and Posttraumatic Growth in Suicide Loss Survivors." Ph.D. dissertation. Philadelphia: Philadelphia College of Osteopathic Medicine. Retrieved 24 April 2023 from https://digitalcommons.pcom.edu/psychology_dissertations/475.

Brewer, John D. 2020. "Remembering Forwards: Healing the Hauntings of the Past." In *Post-Conflict Hauntings: Transforming Memories of Historical Trauma*, ed. Kim Wale, Pumla Gobodo-Madikizela, and Jeffrey Prager, 29–46. Palgrave Studies in Compromise after Conflict. Cham: Palgrave Macmillan. https://doi.org/10.1007/978-3-030-39077-8.

Bruch, Carl, David Jensen, Mikiyasu Nakayama, Jon Unruh, Rebecca Gruby, and Ross Wolfarth. 2009. "Post-Conflict Peace Building and Natural Resources." *Yearbook of International Environmental Law* 19(1): 58–96. https://doi.org/doi.org/10.1093/yiel/19.1.58.

Büssing, Arndt, Andreas Michalsen, Sat Bir S. Khalsa, Shirley Telles, and Karen J. Sherman. 2012. "Effects of Yoga on Mental and Physical Health: A Short Summary of Reviews." *Evidence-Based Complementary and Alternative Medicine* 2012. https://doi.org/10.1155/2012/165410.

Dudouet, Veronique. 2007. *Surviving the Peace: Challenges of War-to-Peace Transitions for Civil Society Organisations.* Berlin: Berghof Research Center for Constructive Conflict Management. Retrieved 24 April 2023 from https://berghof-foundation.org/files/publications/br16e.pdf.

Dushimirimana, Fabien, Vincent Sezibera, and Carl Auerbach. 2014. "Pathways to Resilience in Post Genocide Rwanda." *Intervention* 12(2): 219–30. https://doi.org/10.1097/WTF.0000000000000036.

Ekore, Rabe Ilemona, and Bolatito Lanre-Abass. 2016. "African Cultural Concept of Death and the Idea of Advance Care Directives." *Indian Journal of Palliative Care* 22(4): 369–72. https://doi.org/10.4103/0973-1075.191741.

Ellis, Wendy R., and William H. Dietz. 2017. "A New Framework for Addressing Adverse Childhood and Community Experiences: The Building Community Resilience Model." *Academic Pediatrics* 17(7): S86–93. https://doi.org/10.1016/j.acap.2016.12.011.

Freeman, Kimberly, Kelly Baek, Michelle Ngo, Veronica Kelley, Elaine Karas, Stephanie Citron, and Susanne Montgomery. 2021. "Exploring the Usability of a Community Resiliency Model Approach in a High Need/Low Resourced Traumatized Community." *Community Mental Health Journal* 58: 679–88. https://doi.org/10.1007/s10597-021-00872-z.

Gallagher, Susan Vanzanten. 2002. "'I Want to Say: / Forgive Me': South African Discourse and Forgiveness." *PMLA* 117(2): 303–6. https://www.jstor.org/stable/823277.

Goodman, Rachael D. 2013. "The Transgenerational Trauma and Resilience Genogram." *Counselling Psychology Quarterly* 26(3–4): 386–405. https://doi.org/10.1080/09515070.2013.820172.

Grabbe, Linda, Melinda Higgins, Dorothy Jordan, Lauren Noxsel, Barbara Gibson, and Jordan Murphy. 2020. "The Community Resiliency Model®: A Pilot of an Interception Intervention to Increase the Emotional Self-Regulation of Women in Addiction Treatment." *International Journal of Mental Health and Addiction* 19: 793–808. https://doi.org/10.1007/s11469-019-00189-9.

Habimana, Samuel, Emmanuel Biracyaza, Emmanuel Habumugisha, Evariste Museka, Jean Mutabaruka, and Susanne B. Montgomery. 2021. "Role of Community Resiliency Model Skills Trainings in Trauma Healing among 1994 Tutsi Genocide Survivors in Rwanda." *Psychology Research and Behavior Management* 14(July): 1139–48. https://doi.org/10.2147/PRBM.S319057.

Heldring, Leander. 2021. "The Origins of Violence in Rwanda." *The Review of Economic Studies* 88(2): 730–63. https://doi.org/10.1093/restud/rdaa028.

International Peace Institute. 2013. *Peace, Justice, and Reconciliation in Africa: Opportunities and Challenges in the Fight Against Impunity*. New York: International Peace Institute: The African Union Series. Retrieved 24 April 2023 from https://www.ipinst.org/wp-content/uploads/publications/ipi_epub_peace_justiceafrica2.pdf.

Ionescu, Serban. 2011. "Le Domaine de La Résilience Assistée." In *Traité de Résilience Assistée*, ed. Serban Ionescu, 3–18. Paris: Presses Universitaires de France.

Jansen, Stefan, Ross White, Jemma Hogwood, Angela Jansen, Darius Gishoma, Darius Mukamana, and Annemiek Richters. 2015. "The 'Treatment Gap' in Global Mental Health Reconsidered: Sociotherapy for Collective Trauma in Rwanda." *European Journal of Psychotraumatology* 6: 1–6. https://doi.org/10.3402/ejpt.v6.28706.

Julien-Gauthier, Francine, and Colette Jourdan-Ionescu. 2015. "Résilience Assistée, Réussite Éducative et Réadaptation" [Assisted resilience, educational success, and rehabilitation]. In *De La Résilience Naturelle à La Résilience Assistée*, ed. Serban Ionescu, 1–14. Québec, Canada: Presses Universitaires de France.

Kagoyire, Marie Grâce, and Annemiek Richters. 2018. "'We Are the Memory Representation of Our Parents': Intergenerational Legacies of Genocide among Descendants of Rape Survivors in Rwanda." *International Rehabilitation Council for Torture Victims* 28(3): 30–45. https://doi.org/10.7146/torture.v28i3.111183.

Korman, Rémi. 2016. "When the Body Is Forgotten: The Physical Consequences of Genocide for the Survivors." *Témoigner: Entre Histoire et Mémoire* 122: 150–53. https://doi.org/10.4000/temoigner.4678.

Koulen, Sarah-Jane. 2009. "Book Review: *Traditional Justice and Reconciliation after Violent Conflict: Learning from African Experiences*, ed. Luc Huyse and Mark Salter." *Journal of African Law* 53(2): 321–25. https://doi.org/10.1017/s002185530999009x.

Krog, Antjie. 2015a. "Research into Reconciliation and Forgiveness at the South African Truth and Reconciliation Commission and Homi Bhabha's 'Architecture of the

New.'" *Canadian Journal of Law and Society / Revue Canadienne Droit Et Société* 30(2): 203–17. https://doi.org/10.1017/cls.2015.14.

———. 2015b. "Rethinking Reconciliation and Forgiveness at the South African Truth and Reconciliation Commission." *International Journal of Public Theology* 8(4): 9–23. https://doi.org/10.1163/9789004339569_003.

Leitch, Laurie, and Elaine Miller-Karas. 2009. "A Case for Using Biologically Based Mental Health Intervention in Post-earthquake China: Evaluation of Training in the Trauma Resiliency Model." *International Journal of Emergency Mental Health* 11(4): 221–33. https://www.thresholdglobalworks.com/pdfs/post-earthquake-china.pdf.

Lepore, Stephen, and Tracey Revenson. 2006. "Resilience and Posttraumatic Growth: Recovery, Resistance, and Reconfiguration." In *Handbook of Posttraumatic Growth: Research & Practice*, ed. Lawrence G. Calhoun and Richard G. Tedeschi, 24–46. Mahwah, NJ: Lawrence Erlbaum Associates Publishers.

Luthar, Suniya, and Dante Cicchetti. 2000. "The Construct of Resilience: Implications for Interventions and Social Policies." *Development and Psychopathology* 12(4): 857–85.

Mafeza, Faustin. 2013. "Restoring Relationship between Former Genocide Perpetrators and Survivors of Genocide against Tutsi in Rwanda through Reconciliation Villages." *International Journal of Development and Sustainability* 2(2): 787–98.

McNamee, Terence, and Monde Muyangwa, eds. 2021. *The State of Peacebuilding in Africa: Lessons Learned for Policymakers and Practitioners* Cham: Palgrave Macmillan. https://doi.org/10.1007/978-3-030-46636-7.

Miller-Karas, Elaine. 2015. *Building Resilience to Trauma: The Trauma and Community Resiliency Models*. New York: Routledge.

———. 2018. "Community Resiliency Model (CRM)® Teacher Training Certification." *Trauma Resources Institute*. Retrieved 24 April 2023 from https://www.traumaresourceinstitute.com/materials/building-resilience-to-trauma-second-edition-paperback.

Ogińska-Bulik, Nina, and Magdalena Kobylarczyk. 2016. "Association between Resilience and Posttraumatic Growth in Firefighters: The Role of Stress Appraisal." *International Journal of Occupational Safety and Ergonomics* 22(1): 40–48. https://doi.org/10.1080/10803548.2015.1109372.

Pascoe, Michaela, Alan P. Bailey, Melinda Craike, Tim Carter, Rhiannon Patten, Nigel Stepto, and Alexandra Parker. 2020. "Physical Activity and Exercise in Youth Mental Health Promotion: A Scoping Review." *BMJ Open Sport and Exercise Medicine* 6(1): 1–11. https://doi.org/10.1136/bmjsem-2019-000677.

Patel, Vikram. 2011. "Traditional Healers for Mental Health Care in Africa." *Global Health Action* 4(1). https://doi.org/10.3402/gha.v4i0.7956.

Pfefferbaum, Rose L., Betty Pfefferbaum, Richard L. Van Horn, Barbara R. Neas, and J. Brian Houston. 2013. "Building Community Resilience to Disasters through a Community-Based Intervention: CART Applications." *Journal of Emergency Management* 11(2): 151–59. https://doi.org/10.5055/jem.2013.0134.

Pham, Phuong N., Harvey M. Weinstein, and Timothy Longman. 2004. "Trauma and PTSD Symptoms in Rwanda: Implications for Attitudes toward Justice and Reconciliation." *JAMA* 292(5): 602–12. https://doi.org/10.1001/jama.292.5.602

Rafferty, Rachel. 2017. "Engaging with the Violent Past to Motivate and Direct Conflict Resolution Practice in Northern Ireland." *Conflict Resolution Quarterly* 35(2): 197–221. https://doi.org/10.1002/crq.21204.

RGB (Rwanda Governance Board). 2017. *Impact Assessment of Umuganda 2007–2016.* Kigali, Rwanda: Rwanda Governance Board. Retrieved 24 April 2023 from https://www.rgb.rw/fileadmin/user_upload/RGB/Publications/HOME_GROWN_SOLUTIONS/Impact_Assessment_of_Umuganda_2007-2016.pdf.

Richters, Annemiek, Cora Dekker, and Willem F. Scholte. 2008. "Community Based Sociotherapy in Byumba, Rwanda." *Interventions* 6(2): 100–16. https://doi.org/10.1097/WTF.0b013e328307ed33.

Richters, Annemiek, and Marie Grâce Kagoyire. 2014. "Of Death and Rebirth: Life Histories of Female Genocide Survivors." *Journal on Rehabilitation of Torture Victims and Prevention of Torture* 24 (1 Suppl. 1): 6–11.

Santayana, George. 1905. *The Life of Reason.* New York: Charles Scribner's Sons.

Sorsdahl, Katherine, Dan J. Stein, Anna Grimsrud, Soraya Seedat, Alan J. Flisher, David R. Williams, and Landon Myer. 2009. "Traditional Healers in the Treatment of Common Mental Disorders in South Africa." *The Journal of Nervous and Mental Disease* 197(6): 434–41. https://doi.org/10.1097/NMD.0b013e3181a61dbc.

Tan, Mengxin, Yuko Otake, Teisi Tamming, Valerie Akuredusenge, Beatha Uwinama, and Fabien Hagenimana. 2021. "Local Experience of Using Traditional Medicine in Northern Rwanda: A Qualitative Study." *BMC Complementary Medicine and Therapies* 21: 210. https://doi.org/10.1186/s12906-021-03380-5.

The World Bank. 2023. "Overview." *The World Bank.* Retrieved 6 January 2023 from https://www.worldbank.org/en/country/rwanda/overview.

Uwimbabazi, Penine. 2012. "An Analysis of Umuganda: The Policy and Practice of Community Work in Rwanda." Thesis. Pietermaritzburg, South Africa: University of KwaZulu-Natal.

Wrenn, Glenda L., Aliza P. Wingo, Renee Moore, Tiffany Pelletier, Alisa R. Gutman, Bekh Bradley, and Kerry J. Ressler. 2011. "The Effect of Resilience on Posttraumatic Stress Disorder in Trauma-Exposed Inner-City Primary Care Patients." *Journal of the National Medical Association* 103(7): 560–66. https://doi.org/10.1016/S0027-9684(15)30381-3.

CHAPTER 6

Sankofa

A Model of Sustainable Development

PAMELA WALDRON-MOORE

A little more than two decades ago, a connection was made between climate change and low development. It is acknowledged that the impact of climate change on development has rendered vulnerable societies of the Global South not yet ready to achieve sustainability of their developmental goals. Yet, although the twentieth century has not provided much hope for sustainable development in Africa and many parts of its diaspora, twenty-first-century shifts in an understanding of the causes of climate change, as human-induced, suggest that there is scope for transformation in the Black diaspora from low development to potentially strong and sustainable achievements in that regard. Some of the reasons for hope stem from a recognition that knowledge sharing on the ecological, political, economic, and cultural impacts of development might contribute to more responsible and just human behavior and a greater will to work within communities to develop economies of scale that may be adaptable to the transformations Africa and its diaspora would need in order to envision sustainable levels of development.

As Singapore Minister for the Environment and Water Resources Masagos Zulkifli (2019) eloquently argued in his keynote address to the Intergovernmental Panel for Climate Change (IPCC), the world is witnessing geopolitical shifts, disruptive technological advancements and changing trade patterns that require the global community "to rethink traditional models and practices." He added that despite awareness and concern about the rapid intensification of climate change, governments at either end of the geopolitical spectrum engage in destructive behavior in pursuit of their own national interests. Such destructive and self-interested behavior must give way to a multilateral cooperative framework for change if the existential challenge of climate change is to be adequately addressed to meet the goals of the IPCC and, in the process, achieve sustainable development in vulnerable disaster zones of the Black diaspora.

A geopolitical shift away from national to global collaboration on climate recovery and sustainable development, while an urgent imperative, must first take stock of the range of adaptations needed across countries and the scale of adjustments possible in vulnerable localities such as Haiti, Louisiana, and Rwanda, whose natural and cultural exposure to disaster further complicates the challenge of climate change. To identify root causes of the problem of development in developing nations (largely inclusive of Africa and its diaspora), one has to look to the geographic, political, economic, and cultural history of its people. Looking back in time may confirm the challenges for development in a region where there has been high dependency on an ecosystem that yields a livelihood from agricultural growth, marine sustenance, cattle farming, and human resources. That dependency has taken advantage of the earth's lush resources without fully taking the time to minimize land degradation, treat disease, or reduce the social and political consequences that attend the challenge of survival as populations increase. Replenishing the ecosystem while benefiting from it is a lesson from the past that may have been forgotten in the human urgency to survive underdevelopment and realize growth. Yet, it is in humans' plunder of the ecosystem and inattentiveness to the role each human must play in maintaining it that future generations are tasked with recovery. Responding to climate stress has become a priority for those countries vulnerable to disaster and thus more negatively impacted by climate shifts.

Sankofa Modeling

The *Sankofa* image in figure 6.1 symbolizes the Akan people's belief in a quest for knowledge. In African historiography, Ghanaian people of the Akan culture (see Agbo 1999) are encouraged to engage in critical examination and patient investigation in order to gain the wisdoms of the past and ensure a strong sustainable future built on understandings of what may have succeeded or failed before attempting to implement political, cultural, social, and/or economic change. The objective of this symbol was to empower Africans so that they may not neglect what "was at risk of being left behind" in the passage of time but to "look, seek and take it" into the future. Across the African diaspora, displaced people of Akan heritage and beyond are reminded, by way of the Sankofa symbol (figure 6.1), that the past serves as a guide for planning the future. For an examination as critical at this time as the sustainability of Africa and its diaspora, heeding the symbolism of this entreaty to diasporic communities is a timely re-

Figure 6.1. The Sankofa. Mythical African Bird. Wikimedia Commons, public domain.

minder that humans have degraded the ecosystem and without a return to the wisdoms of the past, they may not be able to restore a viable and sustainable future for generations to come.

This study will proceed then on the understanding that the impact of climate change on development in the diaspora has been devastating. As human behavior continues to impact and degrade the ecosystem, which has been primarily responsible for the sustenance of populations vulnerable to the ravages of climate disaster, it is imperative that action be taken by each society to contribute appropriately to mitigation efforts to create a climate-sustainable future. How may diasporic communities, challenged by limited financial institutions, weak institutional infrastructure, displaced human resources, and heavy dependence on the soil, overlook the political, economic, social, and cultural disasters spawned by climate change? Addressing lessons of history may allow researchers to:

1. begin to understand climate science and the obstacles to development that lie ahead as a result of the adverse climate forecast;
2. critically assess areas of vulnerability relevant to the diaspora's responsibility for upkeep of the ecosystem, as well as the role others have played in the past and must continue to play to adapt responsibly to the needs of the present;
3. identify priorities that may be set in relation to adaptations for sustainable goals to be achieved as recommended by the international community (see UN 2020);
4. develop strategic plans for the implementation of adaptation measures now urgently required to mitigate disaster as climate change heightens;

5. generate knowledge, education, and training, and make accessible to all citizens, critical information on the work of agronomists and climate scientists over time so that knowledge sharing can become the springboard from which recovery may emanate;
6. be cognizant of models and strategies that worked in the past to ensure cultural growth, which can be replicated to the scale currently needed and be mobile enough to reach deep into all communities of the diaspora;
7. be mindful of both the challenges and opportunities embedded in the implementation of new practices aimed at overcoming the harsh realities of climate change and the need for mitigating disaster in the interest of attaining sustainable development;
8. and, most importantly, be open to knowledge sharing in collaboration with all sectors of the community—public and private—working from bottom-up rather than top-down methods that are inclusive and primed for excellence in achievements.

This chapter will address critical issues for development and climate changes as laid out by the IPCC and guided by the UN Goals for Sustainable Development from a trans-diasporic lens and from a multidisciplinary perspective, reviewing the cultural wisdoms of the past while adapting to the requirements of the present. Black communities within the developed and developing world will be considered to highlight the necessity for equity in an approach to global climate change and sustainable development. The areas primarily targeted in this study are areas of underdevelopment that have been recognized globally as risk factors for developing nations and the Black diaspora. Gender inequality, environmental injustice, and economic insecurity, all included in the seventeen goals highlighted in the UN Goals for Sustainable Development, are the primary lenses through which this volume observes the potential impact of knowledge production for achieving sustainability in disadvantaged and disaster-prone areas of the African diaspora.

Sustainable Development Goals

The impact of climate change is felt across the globe, varying among regions, generations, age, class, income groups, educational levels, and gender. Based on the adaptations recommended by the Intergovernmental Panel on Climate Change (IPCC), it is clear that people who are already very vulnerable and most marginalized are disproportionately impacted and need innovative strategies to prevail. The poor, primarily in developing societies, and further burdened by catastrophic weather, are more likely to be in the greatest need of new practices to adapt to climate variability and change. Communities in the diaspora, whether located in the Global North or Global South, many struggling with coastal erosion and infrastructural failure, experience the injustices of disaster management and recognize the physical, social, cultural, and political vulnerabilities that attend them.

The impact of disaster experiences on Black lives complicates the struggle for development. With the evolution of social movements and the realization that the Black experience is one of underdevelopment, subpar growth, and a record of low performance across the spaces inhabited by Black residents, there is no more pertinent time than now to comprehend the implications of injustice and the barriers to development as itself a global disaster rooted in the cultural history of the diaspora and requiring a Sankofa model for change.

The United Nations has laid out seventeen goals for sustainable development. Tackling these goals will help us imagine an ecosystem and a humanity that is equitably shared by those who inhabit the planet. Four of the seventeen goals espoused by the United Nations general body are a priority for study in the context of the Black diaspora. They are Goal 1: to end poverty in all its forms everywhere; Goal 5: to commit to the achievement of gender equality and to empower all women and girls; Goal 6: to ensure availability and sustainable management of water and sanitation for all; and Goal 13: to take urgent action to combat climate change and its impacts.

Goal 1: End to Poverty

Current economic data for the developing world are hard to come by and from the annals of history, including economic archival data, little has changed in that respect. When the UN's Sustainable Development Goals Report of 2020 entreated nations to do everything in their power to end poverty across the globe, there was an implicit objective of fairer distribution of the world's wealth. Although many scholars have debated the connection between global poverty and international distributive justice (see Fuller 2005; Nagel 2005; Olson 1983; Pogge 2001; Singer 1972; Wenar 2003), their scholarship has predominantly focused on the problems of inequality and avenues for addressing them. Little specific attention is focused on who is experiencing poverty versus who has the capacity to identify solutions to the problem and resources to implement actionable plans. Peter Singer (1972), for example, advances the argument that if it is in a nation's power to help without sacrificing anything of moral importance to do so, then that nation ought morally to help. Thomas Pogge (2001) was more targeted in his perspective of how the question of global poverty might be solved. He considered the question from a rationale of resource cost, highlighting the fact that everyone is responsible for the earth's resources, although the Global North contributes disproportionately to the degradation of the environment while perpetuating affluent lifestyles. He proposes that the wealthy West should be taxed on the natural resources they use and that the proceeds of this tax be used to ameliorate the condition of the poor. Providing improvement of the condition of the world's poor by utilizing the dividends of such a tax would not require any sacrifice from the world's wealthy and therefore would be more morally just as well as practically feasible.

The discourse on distributive justice has been extended over the decades beyond the question of morality and feasible practices to the more intractable nature of the problem: that is, how could global poverty be alleviated and what strategies might be put in place to ensure that equity for all sharing the planet might be realized. Global poverty has not decelerated while the world figures out best practices for alleviating the large gap in human suffering. Affluent nations have avoided taking responsibility for environmental degradation and have instead pointed fingers at governments in societies where the greatest disparity in wealth might be found. The latter are often identified as non-democratic, non-transparent, illegitimate, and so forth, and several challenge the leadership of "failing" states as the cause of poverty without addressing the extractive practices imposed on the natural resources of those states, which have led their publics to question the meaning of sovereignty and citizenship bestowed upon their relatively new politically independent state. Lacking the right to self-determination and basic recognition of human dignity and civil rights have left many of the world's poor merely to imagine a world where justice is at the heart of interactions with foreign governments, their own governments, and their interactions with each other. The challenge to end poverty in Haiti, an independent Republic, or Louisiana, where a Black diaspora is located on the southern coast of the wealthy United States, or Rwanda, a relic of European cultural malpractices that created division among the indigenous cultural groups of that land, still begs the question who will bell the cat—a tale in Caribbean folklore.

The Black diaspora has had a front-row seat to squabbles between the Global North and the Global South around the question of distributive justice. Post-colonial societies have debated the value of reparations as a settlement for the inequalities arising from slavery, the disparities observed in migration patterns, and the willingness of global actors and international institutions to acknowledge that inequalities across the globe

Table 6.1. Race-Gender Wage Comparisons in Louisiana. © Pamela Waldron-Moore.

State	Local Average	National Average	Race-Gender Comparison
Louisiana	19% in poverty	12%	All
	44% female-headed	33.8%	All
	\$0.47 to \$1.00	\$0.62	Black women/White men
	\$0.52 to \$1.00	\$0.54 to \$1.00	Latina women/White men

go beyond material resources to human capital as well. And, more significantly, it extends to consideration of who is on the bottom rung of the inequality chart. A quick glance at the wage gap in Louisiana, where approximately 33 percent of the population is Black, highlights this positionality (see Table 6.1).

It is for this reason that the diaspora needs to ponder the lessons of history, to learn how the resiliency of their ancestors was grounded not in gifts from honorable benefactors but in the ability of the poor to convert what little resources they could claim into what they needed to survive and to prevail. Being on the bottom rung of the economic, political, and social ladder forces knowledgeable people to recognize, as Kimberlé Crenshaw Williams (1995) advocates, that to get away from the margins of society, one must understand the "intersectionality" of social domination in order to disrupt such domination. Whether drawing on the songs of empowerment in the struggle for civil rights, for example, "We Shall Overcome," or the narratives of griots, spinning the tales that hunted lions could not tell because the hunters had already glamorized what really happened in the forest to their unsuspecting prey, or the Caribbean fables of Anansi, the spider, who outwitted the cleverest impostors who would suborn the dignity and intelligence of the indigenous people, the power to transform the lives of peoples across the diaspora lies in the courage to identify ways to bell the cat.

It may be useful here to put the moral of the story of the belling of the cat into present-day perspective. In a nutshell, it builds on the saying, "when the cat is away the mice will play." Invariably, as the fable goes, mice are often surprised at play by the stealthy return of the old cat, who then had an easy time making a meal of the mice in their revelry. A plot hatched by the mice to stop this atrocity to their community envisioned putting a bell around the cat's neck while it slept so that its movement in the mice's direction would be hard to miss with the ringing of the bell around its neck. The plan seemed plausible to all the mice in the community until one mouse asked: Who will bell the cat? Of course, there were no volunteers. Despite the well-laid plan to disable the predator rather than eliminate the danger it posed, implementing that plan was as unrealistic as Lisa Fuller's (2005) proposal that NGOs deliver aid to poor countries in order to secure a transparent and accountable system, that is more likely to yield economic security than the tax dividend system proposed by Thomas Pogge (2001), as an opportunity to reduce the burden on poor countries caused by the wealthy countries' wanton degradation of the shared global environment.

A creation of knowledge economies may become the bell around an economy's neck. It could be a blueprint to jumpstart an end to the poverty disproportionately experienced across the African diaspora. The sooner states adopt collaborative tactics, across their geographical boundaries, to combat the predatory extractive nature of affluent states and their corporations, and engage in regenerative practices aimed at capacity building from the bottom up, while relying less on dependency tactics and external organizations such as NGOs to provide aid (often seen as charitable services), the more likely the diaspora is to experience the benefits of knowledge sharing and a reduction of poverty. Equality continues to be a basic presumption of human dignity (Moellendorf 2009; Sangiovanni 2007). The Black diaspora should recognize in the twenty-first century that dignity and human rights have been denied far too long for them to retain the expectation that they will be included in any form of moral justice that will allow them to share in equalizing opportunities provided by the world's affluent peoples. World data has mapped the trends in international versus global poverty and highlighted the differences seen when states are the focus versus people. Often these data exclude Africa and its diaspora from specific discussions on poverty and justice. There is also a shift in international discussion from data on specific states to general evidence of poverty across peoples. As shown in the chart below, an examination of property ownership pinpoints the disparities across class rather than regional inequities, recognizing that 55 percent of adults in the world are classified among the miserable poor, earning approximately less than $10, 000 per annum, which is roughly a 1 percent share of global wealth. Compare this with the 1 percent of adults, who are classified as millionaires, sharing 46 percent of the global wealth. This economic disparity is enough to discourage any group, let alone members of the Black diaspora, from daring to imagine a world where equity and fairness would allow them to benefit from the earth's resources and therefore entitle them to share the environmental costs of sustainable development. Thus, the diaspora's only recourse

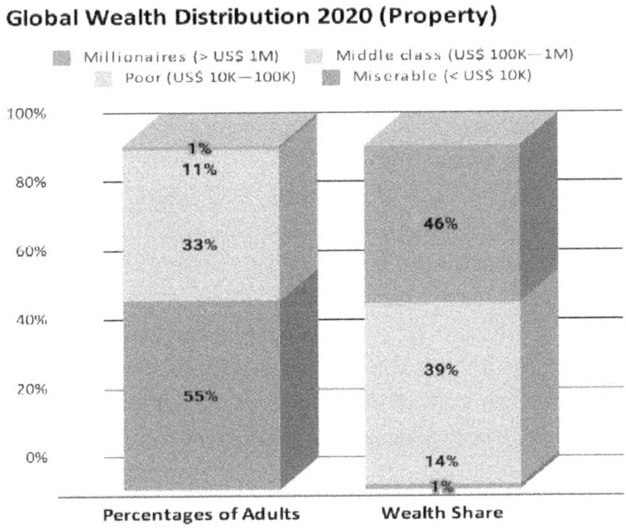

Figure 6.2. Global Wealth Distribution (Property) 2020 Data Source: Credit Suisse Research Institute, Global wealth report 2021, June 2021. Chart created by Leandro Salvador for public domain: https://en.wikipedia.org/wiki/File:Global_Wealth_Distribution_2020_(Property).svg.

would be to forge alliances among themselves by identifying where their intellectual capacities lie, how they might be harnessed and produced through intercommunity efforts to achieve a level of sustainable development that prepares the way for future generations to survive. This cannot be done by replicating the strategies of the West but by pooling the wisdoms of the past and sharing current best practices through collaboration with educators, technologists, engineers, indigenous groups, and more to craft a future where trust and community reliance may yield openness and knowledge sharing for a better collective future, especially among disaster-prone communities, where the urgency of ending poverty and climate mitigation is palpable.

The connection between an end to poverty, on the one hand, and improvements in its correlates (low development, risky infrastructure, climate change, etc.) is inescapable. Developing green infrastructure is at least one method by which poverty may be reduced and disaster recovery from climate change addressed to sustain development in diasporic communities. As recent architectural practices by the nonprofit organization Model of Architecture Serving Society (MASS) demonstrated, its work designing the Butaro Hospital in Rwanda provides a viable opportunity for poverty reduction while providing local jobs and supporting climate change has been a project example of how low development can be elevated in post-disaster recovery. Alan Ricks (2021), the principal architect of MASS, demonstrated how green infrastructure allows for better air flow in hospitals and public buildings to enhance development in poor communities. Rick's experience working in many countries, including some in Africa and the Caribbean, presents an opportunity to observe how collaborative work between nonprofit organizations, engineers, film-makers, healthcare personnel, artists, designers, and landscape architects may not only help empower local activists, builders, and conservationists but also serve as a research tool where lessons learned in countries like Rwanda can be used to teach and educate developed and developing countries on models of architectural design that take into consideration who is involved, who contributes, who benefits, what impacts there are for the climate, and what the value of the design is. Using the cultural wisdoms of Rwanda, architects have developed new models of green infrastructure, thinking of how to build, who should build, where to build, and new innovative ways of building (Ricks 2021). Now it is Rwanda's turn not only to digest what their indigenous environment has taught foreign workers but also how best to share adaptive measures with their diaspora, so that a dent, however small, may be made in the check list of goals for sustainable development.

Embracing the UN's call for ending poverty has been challenging in many small countries, not only because of the difficulty in financing solutions to overcoming poverty but also because, it is often the very strategies to escape poverty that threaten the environment that has led states into poverty in the first place. The case of Guyana, a small developing nation, facing the existential threat of climate change provides the interesting dilemma of meeting UN goals for sustainable development and participating in a global effort to reduce risk while collaborating with the Exxon Corporation to realize a better life by extracting oil from its offshore repositories. In this dilemma, Guyana finds itself in the unenviable situation of losing its last resort at claiming wealth from its natural resources to embracing a fossil fuel crisis that the world needs to reduce to preserve a healthy planet for future generations (Arsenault 2021). It seems a cruel blow to be within grasp of billions of dollars of natural resources and an opportunity to elevate the tiny country from near abject poverty to a state of well-being that will/could arise from the billions of dollars Exxon has promised to share from the oil reservoirs it anticipates extracting at the same time that the IPCC is calling for a reduction of 2 percent of current greenhouse gas levels. How could this diasporic locale seek to end poverty and lower carbon emission levels at the same time? Such existential crises put the UN goals for sustainable development and climate protections for global sustainability into stark relief, highlighting the challenges for the diaspora as insurmountable. These are issues that may not be judicially managed in an anarchic world where just transitions are well-nigh impossible to mandate. Identifying the knowledge capital of communities in the diaspora, rich in lessons from the past and intimate experience with capacities of the present may provide opportunities for sustainable futures.

Goal 5: Gender Equality

The United Nations' espousal of a goal that will sustain development across countries is one that can be easily embraced by Black diasporic peoples. Perhaps, there is no idea more attainable to peoples of African descent as that of the role and power of women to lead and build communities that are sustainable. Feminist theories propagate the belief in full social, economic, and political equality for women. The United Nations has called on the international community to commit to the achievement of gender equality and to empower all women and girls to reclaim their humanity from the wanton abuse and violence young women experience via human trafficking and the slate of ills accompanying it. International Relations theories further propose that gender equity is a human right with critical consequences for global development (see Kegley 2009). Indeed, some scholars go so far as to find that "robust democracy is extremely rare in societies that marginalize women" (see Coleman 2005). In addition, as the philosopher Herbert Spencer (1820–1903) argued, a people's condition may be judged by the treatment women receive under it. This is a philosophy still cited in the twenty-first century (see Francis 2007).

At the United Nations fourth conference on women, there were cultural arguments for/against gender equity, with consensus emerging that gender empowerment should be based on the conviction that only the realization of the full potential of all human beings can contribute to true human development. Much scholarship has been invested in an attempt to sensitize the world to the neglect of gendered issues and the place of women in global society. This discourse aims at identifying a theoretical vision that empowers women, secures their basic human rights, and challenges realist theories that the power and development of a state lies in its ability to defend the state when borders and ideals are challenged (Hunt and Posa, 2001; Tickner 2003) rather than concern itself with the protection of those too weak to defend borders. Again, with the state being the dominant focus in global political discourse, it is not surprising that people are stratified in terms of their worth to the state.

In feminist political thought, gender does not stand alone. Race and gender often go hand in hand and need to be addressed in specificity. Critical race theory addresses this notion by arguing that societies and their laws often overlook the fact that race and gender are interpolated under the law and that women of the Black diaspora endure race-ing and gender-ing (Crenshaw Williams 1995) to their disadvantage within societies, such that laws become discriminatory and add an extra dimension to what might be sustainably achieved. In the context of development, then, one must consider development in the Black diaspora not just an equity issue but as social domination in an intersectional context that further places sustainability out of reach of diasporic peoples in Africa, the Caribbean, and the Americas. In a recent African Caribbean Pacific (EU-ACP) project in collaboration with the European Union, sub-indexes show that the region has closed 69 percent of its overall gender gap and is performing well on educational attainment and in the health and survival of its citizens (Cabrera 2000).

Theories of knowledge support the view that access to education, when widely available, prepares publics for self-discovery, leadership, and most importantly, self-fulfillment. What locations in the diaspora have done successfully, for example, Rwanda, Jamaica, Trinidad and Tobago, Barbados, in allocating space for women to thrive as parliamentarians and create policy avenues for other women to chart their own development, is precisely what the diaspora can share in order for each region to design a pathway through innovative exploration to sustainable development.

A commitment to Goal 5 of the United Nation's sustainable development goals has been both a global and national challenge that has made the search for solutions particularly difficult since each society needs to create idiosyncratic plans for its own self-determination based on its own cultural wisdoms and to its own political, social, and economic investment scale. Because states in Africa and its diaspora have been limited in their capacity to control their own political and economic affairs, being reliant upon external economies that hold them hostage to predatory market agreements and lending opportunities, they have tended to cling to the status quo and neglect taking time to consider their collective intellectual capital and how it might be harnessed in search of new methodologies for growth. Gender roles can be redirected to become political realignments, not with former colonial leaders but with regional innovators and collaborators. Pooling those resources and learning from the progress of sisterly nations can empower communities to resolve challenges to recovery in post-disaster environments.

While smaller diasporic communities in the Caribbean and Americas have managed to forge some cultural unity, there are others that will continue to struggle even when embraced by supportive sister-nations.

Haiti finds itself caught in the crosshairs of Caribbean unity. For while Haiti has joined the Caribbean Economic Community (CARICOM) and is being supported via economic knowledge sharing, there are several identity features that isolate it from the English-speaking Caribbean. Language is one such characteristic, although communication is not an insurmountable obstacle. Francophone and Anglophone culture undergirds value conflicts, a product of enslavement that is still embedded in the Caribbean psyche. It is often difficult to navigate between a "divide and conquer" political value system and an "assimilationist" appropriation of values.

Caribbean countries have a record of strong participation of women in politics. This has given women access to decision-making forums. In the twenty-first century, several Caribbean states have succeeded in electing female heads of government, for example, Prime Minister Mia Mottley of Barbados (elected in 2018), Prime Minister Portia Simpson Miller of Jamaica (2012–2016), Prime Minister Kamla Persad-Bissessar of Trinidad and Tobago (2010–2015), and the late Prime Minister Eugenia Charles of Dominica, who served from 1980 to 1995. There has also been an increase in women's representation in politics in countries such as Guyana and Grenada, where there is over 30 percent representation in the House of Representatives. In the Dominican Republic, the only country apart from Rwanda

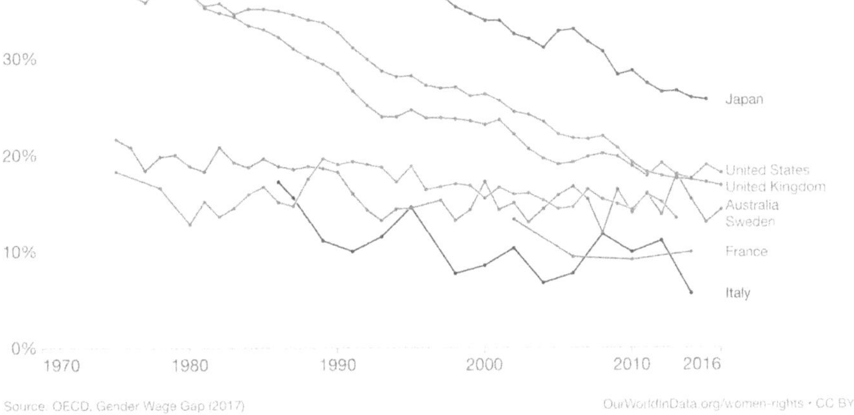

Figure 6.3. Global North Gender Gap, 1970-2016. Source: https://ourworldindata.org/grapher/gender-wage-gap-oecd?country=AUS~FRA~ITA~JPN~SWE~GBR~USA.

that adopted a quota law, women occupy 10 percent of the Senate seats and 21 percent of the seats in the House of Representatives, giving them some agency over policy decisions. A high level of educational attainment has been maintained in the Caribbean diaspora. Thus, women have been able to enjoy a large share of employment in the professional sectors of the region. Most of their engagement, however, has been restricted to the service sectors and to the nurturing occupations (teaching, nursing, social work, etc.), which keep their earning power fairly low. Although via CARICOM access, the African diaspora in the Caribbean has minimized the gender gap over the last three decades, not enough has been done at the grassroots level to promote advancement for poor communities.

While the Black diaspora shares some of the similarities found in explanations of the global gender gap (as illustrated in figure 6.3 above) and may get a thumbs-up for its record on participation in decision-making, there are still areas of performance that may aid those localities in achieving the sustainable goals set by the international community. For example, in recent analysis, it was observed that as younger women gain more access to decision-making, opportunities to intersect voice (in decision-making) and agency (in spending power) may translate into further closing the gender gap globally. This analysis further demonstrated that approximately 21 percent of the variance in household income is explained by younger women (aged fifteen to forty-nine) making decisions about the purchase of major household items. In South America, for example, about 80 percent of women making such decisions spend in the area of $8,800 on household purchases. Comparatively, women in the North American region spend about twice as much and women in East, Central, and West Africa spend about $35,500. There may, of course, be multiple explanations for this, one such being the gap in need as it relates to what items are easily available in the Global North versus the Global South. But, the budgeting decisions of women and potential expenditure on products in the global economy should not be disregarded in conversations on the role such spending may have on the economic well-being of a region. In fact, making financial choices that impact the global economy and utilize expenditure of foreign currency is a resource that merits collaborative attention primarily to ensure that such expenditure by households is geared to the common good and can be redirected toward the growth of innovative homegrown entrepreneurship rather than further importation of goods for which quality alternatives may be found within the native regions. This is a lesson that some women have already learned and implemented in the mitigation of economic disaster at the local level.

Across the United States and specifically in Louisiana, the leadership of women in spearheading innovative projects is astounding. At a presentation on disaster management on 15 July 2021, Dr. Aubrey Paris, policy advisor to the US State Department on global women's issues, moderated a forum entitled *Innovation Station of the Gulf Coast* featuring seven women invested in innovative programs in the Gulf of Mexico and as far afield as Hawaii and Argentina (S/GWI 2021–2023). The panels' conversations were informative, comprehensive, and abundant in their innovative resources. In Louisiana, the Water Collaborative highlighted a diverse arena of project foci, including urban conservancy, landscape architecture, art-centric culture aimed at enhancing the culture of Louisiana, while educating the populace on shoreline resilience, water management, and efforts to deepen the literacy of students and adults in the urban and rural communities of Louisiana. Similarly, challenges for shoreline resilience were tackled by Laura Bowie of the Mississippi Gulf Coast community project working on coastal resilience and green infrastructure in collaboration with communities in neighboring Louisiana. Knowledge sharing on such strategies with other disaster-prone areas of the African diaspora, such as Haiti, Puerto Rico, St. Maarten, and the Bahamas, whose collaboration at the community level with those knowledgeable in climate and infrastructure science, innovative technologies, education, and training in issues pertinent to development, will help communities to harness their capacities for sustainable development.

For the islands of the Caribbean, much benefit could be gained from studying the innovative strategies undertaken by Monterey Seafood Watch (Kemmerley 2015). Sadly lacking in Caribbean communities are neighborhood innovations pertaining to clean-up and financing of community projects for development and sustainability of their coastal areas. As was shared by Aleksandra Dragozet (Netherlands), Karina Campos (Argentina), Helena Williams (Louisiana), and Jeanette Gurung (Hawaii), panelists on the 2021 *Innovation Station* program, efforts to combat climate change, manage issues of food security, educate communities on how to conduct regenerative farming, and help women find their feet in the contemporary world of human resources and sustainable development are likely to rest on community leaders and their commitment to creating sustainable environments for their families, as the realities of climate change bring further devastation to poor societies with low opportunity for sustaining development.

Not only is the innovative work of women in vulnerable societies a powerful reminder of women's capacities, their gathering, leadership, and sharing within their communities ought to remind diasporic communities of their histories of exclusion and ways in which they renegotiated behaviors of the past and its legacies. In early post-colonial times, women in Africa were purposefully restricted from cities by local statutes under the dynamics of systems like apartheid or, in the diaspora, where they experienced difficulty finding housing and employment (Mikell 1997). As Kenneth Little (1973) perhaps jokingly remarked, when explaining female urban migration in West Africa, "men followed the money, and women followed the men" (1973: 17), thereby recognizing the powerful, even if subordinate role, women played in creating progress. Regardless, data on women leading activist organizations, staging demonstrations, lobbying for marriage and other benefits, sharing knowledge and innovating change in cities of Africa as well as in the Caribbean and the Americas abound. In Kenya, for example, women have been establishing nonprofit organizations to provide easy access to safe public transportation, thus connecting mobility issues with climate risks. In Naomi Mwaura's TED talk (2021) on Kenya's successful program in gender sensitivity where women are not only given the opportunity to operate their own transportation services but also educated in the language of climate change and the importance of a just transition away from public transportation to personal ownership of mobility systems in the region. Innovative initiatives, such as these, give rise to the recognition that women are not sidelined or kept away from forums of education because they are perceived to be less intelligent or less able; they are suppressed because men reject or fear women's capacity to lead, to budget, to think, to develop competently/efficiently and empower generations of younger women to seek access to the benches of power in order to better manage state affairs and maintain an active and equal voice in the decision-making process of their communities.

Looking back at examples of the past, Rwanda presents a classic tribute to women's potential for nation-building following the genocide in 1994. That catastrophe left most of the Rwandan men slaughtered by each other in the massacre, allowing a strategic plan to be crafted to integrate women into the work of government. An oversample of women was designed to represent the country in the chambers of government. In 2003, 64 percent of seats in Rwanda's lower house of the national legislature were occupied by women. As a comparison, only 19 percent in the US House of Representatives is held by women. Indeed, only Rwanda and Bolivia have a majority of women assisting in policy-making in their countries. But, does this fact hold any currency for Rwanda in the determination of "governance" or gender equity?

In a report filed by National Public Radio journalist Gregory Warner (2016) affording women space to represent Rwandan publics did not automatically mean for Rwandan women a closing of the gender gap. While in numerical assessment Rwanda ranks sixth among countries narrowing the gender gap in the early twenty-first century, this did not mean that women's roles were modified and that they had greater voice. It simply meant (and continues to mean) that although they would not think of themselves as feminists, in the American/European context, they are attending collectively to nationally ascribed gender roles but standing up, not just for the good of women in the country but for the country as a whole.

The fact remains, too, that most of the Rwandan women who were elected or appointed to serve were not elected on their formal attainment of education (baseline knowledge measure), as in the pre-genocide period but rather on their ability to serve and represent the needs of the public. Whereas prior to the genocide, women did not have the capacity to own their own homes, land, or work in professional occupations outside of the home, appointing women to positions of authority post-genocide, amplified their voices in decision-making and called attention to their natural, intellectual capital. Understanding how their ancestors sacrificed and worked communally to protect their communities, they were more than ready to assume roles they were denied in pre-genocide Rwanda, where external forces were able to generate hostility and dissension, resulting in group violence, rather than curate patterns of development where gender did not propagate exclusivity and divisiveness. The wisdom and resilience women demonstrated post-genocide not only illustrate how closing the gender gap could prove effective for developing the country but also clarify how empowering women to self-actualize, in the absence of a male workforce, for nation-building could open new innovative channels for development and allow women to step in and take their rightful place in building communities for growth. Unlike what happened in the United States and European territories, after men returned from World War I and claimed their traditional roles of running and developing the country and women were forced to return to their domestic duties or to the roles traditionally occupied by them, that is, teachers, nurses, secretaries, and so forth, the women of Rwanda were empowered to successfully produce creative strategies for nation-building, since there was no scope for waiting for the men in their communities to return to claim lost territory.

As the debate on gender equality policies becomes louder in the Black diaspora, scholars have teamed up to explore the meaning of gender equality initiatives in strong male-dominated states (Debusscher and Ansoms 2013). These scholars sought to evaluate deep-rooted societal norms and practices in order to determine factors that inform the level of commitment among Rwandan women to gender equality, the impact of women's commitment to development, and the trends that threaten the potential of Rwandan gender equality to transform gender equity policies. Thus, despite the political will to present target-driven policies and opportunities for promoting gender equality policies, the expected transformative potential was undercut by the dominance of the economic rationale for development already in place; the continued neglect of the contribution of women's labor; the strong focus on quantitative rather than qualitative achievements of implemented policies; the inattentiveness to community voices on issues of policy; and the lack of grassroots participation in the process of policy-making. These and similar findings indicate that in regions where there is scope for gender equality to exist, especially following disasters, patterning leadership on male-dominated, authoritarian values does not allow for genuine gender equity to prevail.

In the contemporary era, Rwandan women have formed communities of developers. Learning from old cultural traditions, women have led the way in the development of textiles and crafts, such as papermaking from banana pulp, which has led to fabric development and paper products. And, building on old cultural wisdoms, women have begun to reclaim the sayings they grew up with, for example (in Kinyarwanda), "knowledge is like a garden: if it is not cultivated, it cannot be harvested." In the post-disaster era since 1994, Rwanda has developed communities of women, some teaching, some learning the entrepreneurial skills that would take them out of poverty and helplessness into leadership, and knowledge production for the development of their communities. The community approach to development has allowed for a remarkable turnaround of fortunes, with women now able to buy land and cultivate it for recovery and growth.

Because of the loss of so many male lives, most of the patriarchy has shifted to shared relationships where young men are being groomed to understand the value of partnerships between men and women and its importance in reforming the culture of resilience rather than hostility and competition as engendered by European occupation and exploitation. Communities have fostered entrepreneurial skills in dress-making, fabric production, and more. They are benefiting from micro-financing collaborations with private and public sector partners and the universities are producing bumper crops of mental health practitioners to provide support for recovery and train Rwandan women in how to be resilient and to make decisions not just for themselves, despite the inequities they have experienced, but for the whole country.

As younger Africans in the diaspora are often reminded by elders in their communities, "the strength of the crocodile is in the water": a reminder that people have their own niches and are strongest in that environment. This serves to discourage unnecessary competition between and among men and women. Thus, one recognizes the subtle ability to identify knowledge and the technologies to convert them into a product that is sustainable. The reiteration of what strategies achieve success and the demonstration of the harmony it brings to think of and work in a space not divided by colonial exploitation but one cognizant of shared values and respect for the dignity and well-being of each member of the community is reinforced. Removing group symbols in Rwanda, for example, aimed at developing hatreds among groups while exploitative strategies to divide and conquer run amuck, was the first step toward Rwandan recovery but establishing equity among communities of males, females, old, young, punisher, punished, and others can today be attributed to Rwanda's successful embrace of the challenges for restoration and opportunities for sustainable development. Such outcomes need to be widely shared across Africa and its diaspora to introduce such community-building to the potential for achieving the goals of the international community to ensure the universe benefits collectively from a potentially sustainable future.

Goal 6: Public Health

UN Goal 6 encapsulates what benefits life and provides a baseline for growth and the pursuit of happiness in a shared ecosystem. To ensure availability and sustainable management of water and sanitation for all is to ensure humanity thrives and regenerates, building on shared experiences for the sustainability of life. Addressing human vulnerabilities across communities will not only allow humans to adapt their behavior to protect the environment from a changing climate but will allow them to apply creative wisdoms for sustainability. Sharing knowledge and expertise in traditional settings helps societies in the diaspora to meet the public health standards identified by the United Nations to claim citizenship in the global commons. Goal 6 is uniquely appropriate to measure innovative strategies in Louisiana, for example, where water and sanitation have featured in every strategic plan groups have developed to redeem patterns of loss, in Louisiana, from disaster as well as from a history of poverty and injustice.

Louisiana, with her large Afro-diasporic population, is easily identified as one of the most disaster-impacted regions of the Global North. Along with other areas of the Gulf South, Louisiana's exposure to hurricanes and ill weather has required, for the assurance of public health, that a safe drinking water program be set in place and that "boil water" advisories and water outages be added to available emergency reports to residents immediately after a storm. The safe drinking water program provides comprehensive mandates, with federal and state regulations, to ensure public health and safety from illness and death that ensures protection to all residents of Louisiana. It monitors public water systems to comply with protective measures against what may be caused by waterborne diseases or contamination of water sources. In the immediate aftermath of a weather catastrophe, the Safe Drinking Water Program activates monitoring exercises aimed at testing for contamination levels and treating such contamination with immediate action. In addition, inspections are conducted and sanitation surveys distributed to determine the capability of the drinking water system to meet the needs of public safety so that, where desired, it can deliver satisfactory quantities and quality of safe drinking water to residential and business consumers. There is also an engineering review plan in place to determine if any new construction or modifications are needed for the system in place. This guarantees that water sources, its treatment, storage, and distribution facilities ensure the quality and protection of drinking water and sanitation necessary for healthy development. Much of this effort targets the urban areas where a majority of Black and brown populations reside.

Marked improvement in Louisiana's water management system has been observed since the Safe Drinking Water Fee (Act 605) was passed in 2016. The collaborative effort among local government officials, federal action committees, members of the private sector, and a variety of public health officials ensure that critical infrastructure, transparent communication, the involvement of technical operators, school boards, political operatives, and opinion leaders are all made accountable for righting the wrongs experienced in public health safety by vulnerable communities of the Black diaspora and indigenous groups resident in Louisiana.

The connection has already been made between climate change and development. Climate change is responsible for increases in disaster frequency that, in turn, lowers already low development, rendering sustainability a struggle for impacted communities. Global leaders assessing performance around the world of water resources reported an audit of said resources in Louisiana. They found, as reiterated at the COC (climate conference) in Glasgow in October 2021, that Louisiana leads the world in efforts to manage the water system. It is noted that public and private investments, supported and sponsored by "world class" science and technology identify Louisiana as a leader to be emulated in its realization of the sustainable development goals of the United Nations as outlined in Goal 6. It is precisely because of a history of loss in a state that is 6 feet below sea level with a large enough diasporic community that this leadership should be emulated by other residents of the Black diaspora. For example, Guyana, also a region below sea level with developmental issues stemming from decades of flooding as a result of overtopped sea walls and reliance on European partners to assist in meeting their developmental goals, could learn from Louisiana's success in managing its water resources.

Interestingly enough, one of the partners to Guyana and states in the Gulf of Mexico is the Netherlands. Dutch assistance in the area of water management has been reinforced by the collaboration of thousands of professionals engaged in the work of water management with a focus on preserving coastal habitats while protecting the millions of residents within/across the Gulf of Mexico. Although Louisiana leaders recognize the benefit of relying on their own cultural wisdoms in exploration of the relationship between water and the survival and sustainability of its people, local leaders are also aware of the Netherlands' efforts in finding solutions to the problems of low-lying areas of Europe and of the balance it has created in efforts to support the coastal challenge created by the confluence of the Rhine, the Meuse, and the Schelde Rivers in the delta. Engaging in a collaborative water management project allows Louisiana not only to learn from Dutch expertise but also to work alongside university professionals, scientists, and technologists across the globe, while hiring thousands of local employees to be part of the management sector of the project.

As research on water proceeds, research tools have been engaged to accurately forecast the flooding impact on neighborhood schools, hospitals, and other structures before a catastrophe strikes. This early warning attention is helpful. It alerts officials to the potential impact of a storm, to the potential damage to the shorelines, to the natural and man-made structures contributing to the devastation of marshes, barrier isles, and communities long held hostage by climatic events such as hurricanes and tropical storms. Findings stemming from such research are critical for other areas experiencing vulnerability from the weather and climate change. This is knowledge that can/should be shared with others in the diaspora. It is understandable that knowledge emanating from Black localities in the Global North can be extremely beneficial to communities in the Global South, who do not have the monetary resources to access such information. Black localities such as New Orleans where most of the services on a water management project are already publicized and shared with residents on the Great Lakes of the Upper Midwest or Vietnam's residents of the Mekong Delta should also be made available to others in the diaspora who can learn and benefit from the research (Ehrenwerth 2017). As Justin Ehrenwerth explained, using science to guide leadership has practical impact. Leading the Water Institute of the Gulf can put science into practice. He thus appreciates the need to take the best available science to develop realistic questions for good policy-making.

The Water Institute of the Gulf of Mexico, under the leadership of Ehrenwerth, serves as an important reminder of the role the past must play in decisions of the present and potential guidance for sustainability in the future. Reflecting on the Sankofa model, introduced here to guide thinking on the role of cultural wisdoms in effecting new trajectories for development in the diaspora, one may recognize that the Dutch have not always been the best stewards of waterways throughout the history of diasporic spaces in states such as St. Maarten, Guyana, Suriname, and others where the Dutch exercised strong influence on water management by marketing their management strategies as solutions for others with similar vulnerabilities, regardless of the cultural disruptions that ensued in the other spaces. Recognizing that many mistakes were made in the building of sea walls and the reliance on structural man-made solutions as opposed to nature-based solutions cautions diasporic peoples against repeating the errors of the past. As disclosed by David Burdick (2021), in a presentation on innovative approaches to climate change and coastal resilience planning, nature-based solutions are found to be more durable for shoreline resistance than man-made structures. In Louisiana, Lake Pontchartrain, a massive man-made lake, is seen to contribute more to the problem than to the solution of past coastal erosion. A reflection on those failures has helped contribute to the work of the Water Institute as well as improved it. Ehrenwerth is the first to admit his understanding of the practical impact of using science to guide leadership of reforms to water management systems. Focusing on what policies should be put in place by collaborating with residents in vulnerable communities, technical experts with knowledge of the cultural wisdoms of the past in challenged communities, understanding the extent of flooding and increases in flooding in communities over the years, and analyzing alternatives to policy in addition to accessing available funds and regulatory parameters, all suggest that engineering skills alone cannot counteract the vulnerabilities of climate change to impacted communities. In the present, being part of a Water Institute, where all contributing parties (ecologists, emergency management officials, engineers, etc.) can engage in successful collaboration, is what would lead to effective practice and, ultimately, sustainable development for post-disaster communities under threat of future harm to the natural environment.

Haiti is ready to learn from the efforts of Louisiana and its emergence as a global leader in climate change in relation to managing water. Haiti has been ravaged not only by hurricanes but also by earthquakes. Its need for developing new green infrastructure to ensure public safety rests also in managing its water sources. Devastated by recent crises in political leadership, Haiti has to find ways to build the kind of collaboration that Louisiana has successfully instituted. Haiti had once been a beacon for economic development in Louisiana. From the introduction of sugar into Louisiana, Haiti demonstrated its ability to share and collaborate with others of the

African diaspora. Now that Louisiana has learned the importance of triangulating business (entrepreneurship), academia, and private and public sector engineering and consulting, this is an opportunity to help Haiti find its feet. And, not much is needed to do so. Access to information is what Haiti requires. The benefits of becoming a global leader in climate change and sustainable development where disaster experience is shared and solutions found is perhaps the simplest step that can be made in the direction of sustainability. Increasing information flow between the two diasporic localities augurs well for change and development. Bridging community leadership and science, both of which abound in this region must start by identifying the intellectual capacity shared within these groups and their potential to learn from each other. Of primary importance to Haiti, however, would be harnessing the political will in Haiti to put class differences aside and work, as Rwandan women do, to salvage development for the whole country in areas where they are most needed, namely, gender equality, environmental justice, and economic security. Getting local scientists back on board, both in areas of the physical and social sciences, will rest on innovative ideas and planning which will go a long way to realizing the UN's Goal 6 of sustainable development. Water and sanitation are the basic elements of a healthy environment. The incidence of hurricanes and earthquakes as well as traumatic political leadership has left only memories of past glory as an incentive for future growth in Haiti.

Goal 13: Combating Climate Change

Goal 13 serves as an umbrella for all to take urgent action to combat climate change and its impacts. Those impacts have been severely felt in the post-disaster diaspora. Louisiana has seen its coast erode and its interior degrade, affecting its livelihood from marine cultivation and its tourist appeal. Its disaster experience has been expanded by the latest catastrophe to impact its way of life. The COVID-19 pandemic has limited its social activities, crippled its nightlife and the hospitality services that attended it, and severely impacted the pocket books of all those relying on the culture of the state. The harm to its built environment has laid bare the poverty that exists behind the hustle and bustle of everyday life. A loss of life, employment, income in periods of weather disaster only expands the devastation that Louisiana communities experience. New perceived threat for the annual hurricane season, where the damage to already vulnerable communities increases displacement as communities repair and recover and exacerbates the urgency for innovative ideas to meet the challenges to development. Where does resilience come from, if not from cultural wisdoms of the past and identifying the capacities to regenerate the present and build sustainable futures?

Similarly, the experiences of Haiti and Rwanda that address the crises of political leadership, challenges to legitimate governance, and the juggling of management criteria to ensure nations do not regress into complacency in the face of disasters past that have threatened employment, income, wellness, and community, all of which are required for sustainable recovery, serve as shareable knowledge capital that may be produced in the search for environmental sustainability.

Adaptation is at the heart of disaster mitigation in the three diasporic societies highlighted in this volume. Insights that contribute to a favorable outcome for the societies in this study may be applied to post-disaster societies seeking a trajectory of development. This emanates from the identification of intellectual capital resident in these societies and from the production of innovative methodologies for potential sustainability. As capital is produced, the achievement of goals will empower communities to generate conversations that will contribute to a feedback loop and regenerate even more innovative ideas for tackling economic insecurity, gender inequality and environmental injustice. To evaluate the impacts of climate change on an already vulnerable and far-flung diaspora, it is necessary to address the incidence of economic insecurities, gender inequity, and climate injustice on development in order to support a just transition for a sustainable future.

Previous chapters of this manuscript have outlined the need for building a resistant green infrastructure that allows adaptation to climate demands. The opportunity to succeed in the building of such infrastructure has costs and benefits. Among the costs is expenditure on education and training for skilled labor, information technology that is easily accessible to all communities within the given state, the identification, financing, and utilizing of local expertise, and the erection of political, economic, social, and cultural, institutional structures to implement innovative ideas and intellectual capacities of enlightened communities, willing to engage locally for the global good. The benefits are multiple also: guided self-reliance, new green jobs, institutional support, forging community partnerships and investment in low-carbon initiatives, and redress of environmental injustice, as illustrated in the table below.

While the above organization of costs and benefits adequately illustrates the impact climate change will have on the whole planet and potential solutions, it must be reiterated that its likely effect on the African diaspora with its wealth limitations, geographical drawbacks with many living on coasts and islands, its history of colonial exploitation, its strong dependence on natural resources for the livelihood of its people,

Table 6.2. Costs, Benefits, and Outcomes of Global Climate © Pamela Waldron-Moore.

Costs	Benefits	Expected Outcomes
Education and training on carbon reduction	Skilled labor force to mitigate risk	New Green Initiatives and monitoring policies: e.g. wind and solar energy
Information technology task forces	Knowledge access; information sharing	Application of models to scale: clean, cost-effective, energy access
New institutional structures	Framework for guided self-reliance	Political, economic, cultural, and social policy advocacy
Loss from predatory business investments	Community investment and cultural engagement	Advocacy for climate justice and empowerment of action for change

its mostly poor infrastructure and fast-growing populations are especially vulnerable to the consequences of climate change. So, there is no neat package of costs/benefits or solutions that will inspire confidence in a quick fix for the diaspora's lunge at sustainable development by the international community's designated 2030 deadline. A city like Lagos, Nigeria, for example, is one of the fastest growing cities in the world. Its population of 15 million people is forecast to double by 2050. This will not only place pressure on the infrastructure but will actually create more costs than benefits.

On the other side of the Atlantic, Guyana, with its three-quarters of a million population dependent on its own abundant natural resources, finds itself in the awkward position of optimizing its newly oil-generated resources at a time when the world needs to reduce its dependency on fossil fuels and adapt to new energy solutions in order to combat the effects of climate change. Sadly, because of brain drain over centuries, a lack of technological expertise, and underdeveloped access to information sharing, both Lagos and Georgetown, find themselves prey to the oil industry moguls of the Global North and have regressed into dependency on the industrialized core countries rather than engage in the policy advocacy and empowerment strategies that knowledge production for development entails. What is even sadder about this scenario is that Guyana and, to some extent, Nigeria both have a history of revolutionary economic leadership in collaboration with the Non-Aligned Movement, the Group of 77, the ACP economic bloc and other international and regional alliances. Lacking, however, is the political will and policy innovations necessary for these and other historically vulnerable nations to break totally free of the shackles with which neocolonialism has poor states bound and gagged.

Sankofa: Symbolism for the Diaspora

Those who suffer the most from the effects of climate change remain the least responsible for its occurrence. Members of the Black diaspora located in the Global North fare better than those in the Global South, despite the fact that, overall, Blacks are rarely counted among the wealthy. There is an inverted relationship between groups in the North and South as it relates to global wealth distribution. Among the world's wealthiest and most developed countries, thirty-seven are members of the Organization of Economic Cooperation and Development (OECD) group. There are, additionally, eighty-one high income countries that are very well developed. The poorest countries, low in development, are approximately seventy-seven in number and mainly located in the South. These are the countries where African-descended people are generally located. The reality that among these people, there is a history of resilience and survival confirms the view that their strength and grit lie in their cultural orientations and their continued survival depends on the will to survive. Many in the developed North claim that climate change is the world's problem and all countries "are in it together." But there are lessons to be learned from anthropologists that remind us to look to our histories for solutions, for learning from the past, lest we repeat its errors. Africa and its diaspora must learn that it is only through collaboration, knowledge sharing, and a resilient spirit that sustainable development is possible.

The quest for knowledge among the Akan is symbolized by the Sankofa emblem. Its implication that a quest for knowledge must be a critical examination, based on patient, intelligent investigation and analysis is a reminder to the diaspora that there is wisdom in learning from the past and that the only way to assure a strong future is to recall the past, understand what succeeded and what failed, before moving on to the introduction of policy that may be foolproof. Many leaders in the African diaspora have depended heavily on their colonial history for guidance at the expense of their African past. There is urgency now to retrieve that past and engage it to its best advantage. As the Akan believed, when one takes the best from the past and brings it into the present, it only augurs well for the future. Ghana's recovery from the depths of underdevelopment to the progress with which they are credited today is a good incentive for areas of the diaspora still struggling under the burden of imperialistic acculturation. Rwanda has recognized that the hatefulness of "divide and conquer" under French/Belgian oppression has contributed to their failures and have taken those lessons to propel themselves into a positive present and a likely sustainable future. Black communities in Louisiana have taken a similar path and returned to the lessons of the past to devise new methods of managing water, the staff of their existence, for the betterment of their daily lives and communities dedicated to sustainability. A look at each of these communities in context amplifies the Sankofa model.

Haiti: Sustainable Goals

The three post-disaster units (Haiti, Louisiana, Rwanda) referenced in this volume as having the potential to provide a blueprint for development in the future, all have powerful histories of loss and struggle, with equal present experiences of resilience and growth, and the courage and hope to press forward in search of a sustainable future. Haiti has had what may only be described as a turbulent political history. Yet, it was the first Black-led entity in the diaspora to reject French colonial rule to become an independent state. That history is marked by the bravery and resilience that the diaspora understands to be a part of its motivation to attain self-determination despite the cost. Although Haiti's present has been plagued by political, natural, and man-made disaster, it still has the glory of its history to sustain it into a new and thriving tomorrow. So, what will it take for a post-disaster Haiti to break free from the interfering prescriptions of the North and neighborly ostracism in the South to reclaim its past glory, resilience, tenacity, grit? Would adopting a knowledge economy along the framework described earlier in this text be a sound recommendation for the achievement of environmental sustainability? Can a return to cultural wisdoms, where griots enlightened communities and generated advocacy of and empowerment for new innovative ideas, fed by strong communication and grassroots action to legitimize development in the interest of the people, while educating new cadres of risk-mitigators to find strategies for development and liberation akin to those conceived by Jean-Jacques Deassalines in 1804 be the catalyst for change in vulnerable Haiti? Beginning with legitimate people-centered governance and built around educational, sociocultural, geopolitical and economic institutions where easy access to information is available and knowledge sharing can revive and empower the Haitian spirit, a new day may dawn in Haiti's ability to protect its environment, and envisage a sustainable carbon-reduced future.

To accomplish such a goal, it may be pertinent to introduce a better way of defining sustainability. As contemporary environmental scholarship (see Abrahams 2014; Goodland 1995; and WCED 1987) suggests, environmental sustainability incorporates in its definition the need for post-disaster activities but also a plan for resources that envisions a just social environment and provides resources to realize such an environment. Thus a modified definition of environmental sustainability created specifically for the post-disaster context must build on past knowledge and include a just representation of sustainability: "Sustainable post-disaster activities provide resources to affected citizens to ensure health and safety and to promote redevelopment, without causing further damage to land or existing structures, exacerbating the impacts of the disaster, or placing undue stress on the natural environment" (WCED 1987).

As Edwidge Danticat illustrates, in her far-reaching insights on Haiti's political, socioeconomic, and cultural realities of life in Haiti, memory is key to Haiti's survival. As Danticat writes (2003), "I know old people, they have great knowledge. I have been taught never to contradict our elders. I am the oldest child. My place is here." The implication that it is the responsibility of generations to pass on knowledge and retain continuity of the histories of the past for the benefit of the future is not lost in Danticat's writings. Indeed, as Maryse Condé's novel (2001) narrates the life of her grandmother, both Haitian writers reveal the importance of historical narrative of elders' lives as harvesting a "rich landscape of memory." Caribbean scholars know only too well that what is missing from the archives on their African ancestors are only to be found in the memories of living, elderly family members, especially grandmothers and great-grandmothers, on/around whose lives their own futures are expected to be built. What Haiti may currently lack in expressions of climate innovativeness, they have excelled in creative expressions of resilience and the Haitian spirit. Whether in the success of books such as Danticat's and other accomplished Haitian writers or in the theatrical productions of artistes such as Leyla McCalla (2021) in her performance of "Breaking the Thermometer to Hide the Fever," the narrative of resilience is repeatedly told. Directed by Kiyoko McCrae in her exploration of the legacy of Radio Haiti-International, the multimedia experience featuring McCalla, Haitian American singer and songwriter, reminds us that the harsh political realities Haitians face and support for journalists who uplifted the voices of Haitian publics experiencing the trauma of their lived realities, express the creative energies of post-disaster Louisiana, Rwanda, and other diasporic spaces as they memorialize the past while expressing the potential for recovery from lived trauma as hope for a sustainable future.

Louisiana: Sustainable Goals

At the 2021 climate change conference—COP26—in Glasgow, Scotland, former US president Barack Obama applauded the work of Louisiana communities fighting on the frontline of climate change. Selected for this study as one of the few northern territories with a significant track record of Black empowerment during relentless social injustice, it is easy to trace Louisiana's history of activism in the face of existential disaster/crisis to the resilience of grassroots community action. Though stories of political and sociocultural activism in the Black diaspora abound, few know the details of why areas like Louisiana, Jamaica, Guyana, and other diasporic spaces have carved out a niche for being the icons they really are. In Louisiana, stories of freedom-fighting women, like the Castle sisters, closely resemble that of Nanny and the maroons in Jamaica, or men like Kofi who led the rebellion of 1763 to overthrow colonial powers in Guyana. The agency that the Castle sisters and co-activists like Dooky Chase claimed in Louisiana (see Dottin-Haley 2020) still resides in historical memory and speaks to the power of a Sankofa model of development. As Blair Dottin-Haley, grandson of Oretha Castle Haley, recounts in his article in the local Louisiana magazine, *64 Parishes*, the Castles are remembered and honored in the naming of communities and creative centers. Their stories and the oral histories of other memorable women and activists of Louisiana, contribute to the legacies of the past, guiding the present and giving insight into the future, in both creative arts and narrative form.

Contemporary advocates of climate justice (e.g., Pichon-Battle 2019) continue the honoring of these icons by active engagement in legal and policy-oriented community projects, in recognition of the spirit of activism on which the memory of Louisiana's heroines, hinges. In their work on climate justice, the Gulf Coast Center for Law and Policy (GCCLP) collaborates with a community of participants for a Green New Deal, linking with grassroots and other climate activists from the region and around the world to keep global temperatures from rising above the 1.5° level. The Gulf South Center for a Green New Deal (GS4GND) is the first regional-led movement for a Green New Deal and is inspiring change across the nation and world. The GS4GND is an interesting movement that builds on long-existing work in the southern region toward climate, racial, and economic justice. It is a useful blueprint for how other areas of the diaspora may build capacity and rely less on external organizations and more on community collaboration.

Other community development projects have fully embraced the adoption of a Sankofa perspective. Entitled the Sankofa Community Development Corporation, this nonprofit community organization in Louisiana hosts a community garden and runs a farmer's market to promote access to healthy food at affordable prices. The effort to counteract food deserts, arising in the disastrous aftermath of Hurricane Katrina, models itself on food production methods of the past along with communal sharing of produce among neighborhood families. The Sankofa CDC is committed to providing "leadership to build, inspire, and serve" (Ferdinand 2022). Additionally, Louisiana boasts a variety of community collaborations whose purpose is primarily to mitigate the risks of disaster and adapt to green infrastructure that promotes the healthy lifestyle of communities while contributing to the sustainable development of the Gulf of Mexico. One such site, appropriately named the Sankofa Nature Trail and Wetland Park, highlights the role of wetlands in absorbing storm water and reducing land subsidence. Wetlands create a habitat for plants and support wildlife, while its trees cool the environment and absorb the rain. Moreover, the synchronized benefits of the wetland project serve to protect the city against the negative impact of hurricane. Working with the indigenous and Black and Brown communities of the region and drawing on the historical connection between humans and the earth has offered many lessons for the African diaspora in reaching back to the past for innovative entrepreneurial strategies that may potentially usher in a sustainable environmental future. Caribbean islands like St. Maarten, Antigua, and Barbuda and several of the smaller states in the Caribbean (Arrindell 2021), which need climate renovation and have encountered climate injustice in terms of what the colonial governments are willing to provide their citizens, could gain valuable traction from the community exchanges and collaborations from which they have historically benefited, though on a smaller scale.

It is also useful to observe that in Louisiana, knowledge sharing comes in all forms. Creative exercises, such as those hosted by the organization No Dream Deferred, have gathered a "cohort of Black and Black Indigenous artists, makers, and creatives who share an interest in building a coalition focusing on decolonizing their creative practices and crafting a shared vision for a more sustainable future" (Turner 2021).

The uniqueness of this creative expression, as directed by Lauren Turner (2021), memorializes the liberation practices of the ancestors' resistance and innovation. The goal is to continue to build and expand the diasporic archive for future generations. Not dissimilar from the productive activities of Rwandan communities in the wake of a catastrophic genocide that forced a wrenching return to roots as communities sought to heal and rebuild, Louisiana activists recognize that the practice of Sankofa is one of remembering and revering the memory of the past. Decolonization of the mind is practiced through ancestral veneration, out of which comes the wisdom for creating innovative and sustainable futures. As Turner (2021) affirms, "liberatory practices for making are not new; we have countless examples within our current archives of our ancestors' resistance and innovation. Our mission becomes to continue to build and expand the archive for future generations. The work we do now sets the tone." Such are the blueprints of development that must be shared widely within the Black diaspora, if communities are to be developed sustainably. The time to learn from the past and organize creative and innovative community activism with recovery in mind is a viable pursuit for a sustainable future.

Rwanda: Sustainable Goals

Rwanda's catastrophe is etched from its colonial past. Its recovery strategies serve as an example for the survival of the diasporic spirit. In the previous chapter, much of Rwanda's efforts to support recovery in the nation illustrated actions taken to honor the past and venerate those who died in the fray and whose corporal remains needed to be disposed of in respectful ways. Similarly, as climate change takes its toll on those seeking to reclaim the soil and replenish the earth, feasible strategies that yielded success in the past should be revised with land-based, plant-based solutions relevant to the topography, landscape, and values of the past. The memorial site in Kigali, not only affords citizens a place to recall the past but also a space where people may reflect on where they have been, where they are and, potentially, where they should be heading. Green infrastructure, relying on alternative sources of energy, clean air, guided water management, and providing wholesome food for healthy living, under the collaboration of a "woke" community and responsible government is a good formula for reducing carbon emissions in an attempt to decrease the warming of the earth and the sustainability of humankind. As is said in the vernacular of Guyanese and others in the English-speaking African diaspora, "one, one, dutty build dam," which suggests that much can be accomplished even a little at a time.

Conclusion: Achieving a Sustainable Future

One fact remains clear in addressing climate change and the diaspora's need to avoid extinction in the face of critical disaster experiences: Waiting for the developed world to be accountable for the harm it has brought to a shared ecosystem and to provide services to the diaspora is not in the interest of a sustainable future. Especially for those threatened by immediate crisis situations, a viable plan must be devised post-haste to divert the mud from hitting the fan. Building knowledge economies, replete with education and training institutions for learning/teaching about climate change and so much more, providing access to information to complement such institutions, identifying intellectual capacity and offering incentives for entrepreneurship and the use of uncovered knowledge capital, and encouraging collaborations with all community stakeholders where innovative approaches can be engaged in public service are foolproof ways of saving communities as they save the planet.

Building alliances across the gender and generational divide is a first step to decolonizing the creative and innovative capitals of communities that have let the narratives of those who claim supremacy over Black thought and action reduce the diaspora to positions of subordination. Barbados's decision to shed the vestiges of colonization by becoming a republic in 2021 is an indication that the diaspora's hope for a sustainable future recognizably lies in renouncing the symbols of oppression of the past and reclaiming a history, fraught with lessons of resilience, innovation, creativity, dignity, and pride. Appointing a woman as the first president of the small island further signals Barbados's readiness to refrain from "loitering on colonial premises" (quoted in Coto 2021). In a passionate acceptance of her new role, Sandra Mason, the first president of Barbados, affirmed that it was time for Barbados to redefine its understanding of self, of state, and the Barbados brand, in a more complex, fractured, and turbulent world where Barbadians need to "dream big dreams and fight to realize them" (quoted in Coto 2021). Less than six months later, Jamaica quietly echoed the sentiment when its Foreign Minister notified the British Crown of its intention to seek Republic status and leave the Commonwealth in the near future.

Together with the Prime Minister of Barbados, the first woman to hold that title, and the new president of the Republic of Barbados, the Black diaspora must recognize that gender inequity, economic insecurity, and environmental injustice can be eliminated. Embracing their knowledge capital, producing said capital, and engaging in regenerative, creative, and innovative practices, is a plausible trajectory for sustainable development. Perhaps, the diaspora, as envisioned by Maya Angelou (1978), can still rise and collectively say:

> Leaving behind nights of terror and fear
> I rise
> Into a daybreak that's wondrously clear
> I rise
> Bringing the gifts that my ancestors gave
> I am the dream and hope of the slave

I rise
I rise
I rise.

Pamela Waldron-Moore, the Leslie R. Jacobs Professor at Xavier University of Louisiana, holds a PhD in political science with specialization in Comparative Politics and International Relations. She has taught at universities in the Caribbean and the United States. Among her research interests are the political economy of development, gendered disparity, and climate justice. She is published in peer-reviewed journals of comparative politics and international studies and contributes regularly to scholarly forums on transformative pedagogy. She is a recipient of Keynote Speaker awards for research on women and the 2018 Jewel Prestage Mentorship Award from the National Conference of Black Political Scientists (NCOBPS).

References

Abrahams, Daniel. 2014. "The Barriers to Environmental Sustainability in Post-Disaster Settings: A Case Study of Transitional Shelter Implementation in Haiti." *Disasters* 38(S1): S25–S49.

Agbo, Adolph H. 1999. *Values of Adinkra Symbols*. Kumasi, Ghana: Ebony Designs and Publication.

Angelou, Maya. 1978. "Still I Rise." In *Phenomenal Woman: Four Poems Celebrating Women*, 7–10. New York: Random House.

Arrindell, Rhoda. 2021. "Disaster Response, Recovery, and Healing in a Colony." *Presentation at the Twentieth Anniversary of the International Conference on Caribbean Literature, themed "Caritharsis: Healing and Regeneration in Caribbean Literature and Culture."* November 11 (virtual). Retrieved 21 April 2023 https://inside.morehouse.edu/events/events-inside/international-conference-on-caribbean-literature.html.

Arsenault, Chris. 2021. "With Oil Boom, Guyana Walks a Tightrope on Growth and Climate." *Al Jazeera Media Network*, 12 November. Retrieved 21 April 2023 from https://www.aljazeera.com/economy/2021/11/12/with-oil-boom-guyana-walks-a-tightrope-on-growth-and-climate.

Burdick, David. 2021. Panel Presentation on "Green and Nature-Based Infrastructure Solutions for Economic Recovery in New England." *American Association of Atmospheric Sciences*. Webinar, October 27.

Cabrera, P. Muñoz. 2000. *ACP Women: Actors of Development*. Brussels: EU/ACP Publication.

Campos, Katrina, Allison Dragozet, Jennifer Gurung, and Heather Williams. 2021. "Women and Innovative Approaches to Climate Change." In *S/GWI's Innovation Station*, podcast moderated by Dr. Aubrey Paris, 15 July. Secretary's Office of Global Women's Issues, US Department of State. Retrieved 21 April 2023 from https://podcasts.apple.com/us/podcast/s-gwis-innovation-station/id1599344045.

Coleman, Isobel. 2005. "The Payoff from Women's Rights." In *Developing World 05/06*, ed. Robert J. Griffiths, 191–96. Dubuque, IA: McGraw Hill.
Condé, Maryse. 2001. *Tales from the Heart: True Stories from My Childhood*. New York: Soho Press.
Coto, Dánica. 2021. "Barbados Becomes a Republic after Bidding Farewell to British Monarchy." *PBS Newshour*, 30 November. Retrieved 21 April 2023 from https://www.pbs.org/newshour/world/barbados-becomes-a-republic-after-bidding-farewell-to-british-monarchy.
Crenshaw, Kimberlé Williams. 1995. "Mapping the Margins: Intersectionality, Identity Politics, and Violence against Women of Color." In *Critical Race Theory: The Key Writings that Formed the Movement*, ed. Kimberlé Crenshaw, Neil Gotanda, Gary Peller, and Kendall Thomas, 357–83. New York: The New Press.
Danticat, Edwidge. 2003. *Breath, Eyes, Memory*. New York: Soho Press.
Debusscher, Petra, and An Ansoms. 2013. "Gender Equality Policies in Rwanda: Public Relations or Real Transformations?" *Development and Change* 44(5): 1111–34
Dottin-Haley, Blair. 2020. "Mama Oretha and Tee Jean: Family Memories of Two Civil Rights Icons." *64 Parishes*. Retrieved 21 April 2023 from https://64parishes.org/mama-oretha-and-tee-jean.
Ehrenwerth, Justin. 2017. "The Water Institute of the Gulf Put Science into Practice." *Young Presidents Organization*, 10 December. Retrieved 21 April 2023 from https://www.ypo.org/2019/12/justin-ehrenwerth-and-the-water-institute-of-the-gulf-put-science-into-practice/.
Ferdinand, Rashida. 2022. "Sankofa Community Development Corporation Impact Report." New Orleans, Louisiana. *Propeller Newsletter: A Force for Social Innovation*, vol. 1(1).
Francis, Mark. 2007. *Herbert Spencer and the Invention of Modern Life*. Ithaca, NY: Cornell University Press.
Fuller, Lisa. 2005. "Poverty Relief, Global Institutions, and the Problem of Compliance." *The Journal of Moral Philosophy* 2(3): 285–97.
Goodland, R. 1995. "The Concept of Environmental Sustainability." *Annual Review of Ecology and Systematics* 26(1): 1–24.
Hunt, Swanee, and Cristina Posa. 2001. "Women Waging Peace: Inclusive Security." *Foreign Policy* 124: 38–47.
Kegley, Jr., Charles W. 2009. *World Politics: Trend and Transformation*. Belmont, CA: Cengage.
Kemmerly, Jennifer. 2015. "Jennifer Dianto Kemmerly, Vice President of Global Ocean Initiatives." *Monterey Bay Aquarium Seafood Watch*. Retrieved 21 April 2023 from https://www.seafoodwatch.org/about-us/mission-and-leadership/jennifer-dianto-kemmerly.
Little, Kenneth. 1973. *African Women in Towns: An Aspect of Africa's Social Revolution*. Boston: Cambridge University Press.
McCalla, Leyla. 2021. *Breaking the Thermometer to Hide the Fever*, dir. Kiyoko McCrae. *Contemporary Arts Center New Orleans*, November. Retrieved 21 April 2023 from https://cacno.org/performing-arts/breaking-the-thermometer.
Mikell, Gwendolyn. 1997. *African Feminisms: The Politics of Survival in Sub-Saharan Africa*. Philadelphia: University of Pennsylvania
Moellendorf, Darrel. 2009. "Justice and the Assignment of the Intergenerational Costs of Climate Change." *Journal of Social Philosophy* 40(2): 204–24

Mwaura, Naomi. 2021. "A Feminist Reimagining of Kenya's Public Transport." *TED Fellows*, 9 June. Retrieved 21 April 2023 from https://www.ted.com/talks/naomi_mwaura_a_feminist_reimagining_of_kenya_s_public_transport?language=en

Nagel, Thomas. 2005. "The Problem of Global Justice." *Philosophy and Public Affairs* 33(2): 113–47.

Olson, Mancur. 1983. "A Less Ideological Way of Deciding How Much Should Be Given to the Poor." *Daedalus* 112(4): 217–36.

Pichon-Battle, Colette. 2019. "Climate Change Will Displace Millions: Here's How We Prepare." *TEDWomen*. Retrieved 21 April 2023 from https://www.ted.com/talks/colette_pichon_battle_climate_change_will_displace_millions_here_s_how_we_prepare.

Pogge, Thomas. 2001. "Eradicating Systemic Poverty: Brief for a Global Resources Dividend." *The Journal of Human Development* 2(1): 59–77.

Ricks, Alan. 2021. "The MASS Model of Community-Focused Architecture." Interview with Lesley Stahl. *60 Minutes*, CBS, 31 October. Retrieved 21 April 2023 from https://www.cbsnews.com/news/model-architecture-serving-society-60-minutes-2021-10-31/.

Sangiovanni, Andrea. 2007. "Global Justice, Reciprocity, and the State." *Philosophy and Public Affairs* 35(1): 3–39.

Secretary's Office of Global Women's Issues, US Department of State (S/GWI). 2021–2023. *S/GWI's Innovation Station*. Retrieved 21 April 2023 from https://podcasts.apple.com/us/podcast/s-gwis-innovation-station/id1599344045.

Singer, Peter. 1972. "Famine, Affluence, and Morality." *Philosophy and Public Affairs* 1(3): 229–43.

Tickner, J. Ann. 2003. "Feminist Perspectives on 9/11." *International Studies Perspectives* 3(4): 333–50.

Turner, Lauren. 2021. "Lauren E. Turner and No Dream Deferred: Building Our Own Tables." Interview with Yura Sapi. *Howlround Theatre Commons*, 16 December. Retrieved 21 April 2023 from https://howlround.com/lauren-e-turner-and-no-dream-deferred.

UN. 2020. "The 17 Goals." United Nations Department of Economic and Social Affairs, Sustainable Development. Retrieved 18 April 2023 from https://sdgs.un.org/goals.

Warner, Gregory. 2016. "It's the No. 1 Country for Women in Politics: But Not in Daily Life." NPR, 29 July. Retrieved 21 April 2023 from https://www.npr.org/sections/goatsandsoda/2016/07/29/487360094/invisibilia-no-one-thought-this-all-womans-debate-team-could-crush-it.

WCED (World Commission on Environment and Development). 1987. *Report of the World Commission on Environment and Development: Our Common Future (Brundtland Report)*. Retrieved 21 April 2023 from https://sustainabledevelopment.un.org/content/documents/5987our-common-future.pdf.

Wenar, Leif. 2003. "What We Owe to Distant Others." *Philosophy and Economics* 2: 283–304.

Zulkifli, Masagos. 2019. "Committee of Supply Debate 2019: Mr Masagos Zulkifli, Minister for the Environment and Water Resources." *Ministry of Sustainability and the Environment, Singapore*, 7 March. Retrieved 21 April 2023 from https://www.mse.gov.sg/resource-room/category/2019-03-07-speech-at-the-committee-of-supply-debate/.

CHAPTER 7

Epilogue

Post-disaster Solutions for Development

PAMELA WALDRON-MOORE

When Audre Lorde (2018) entreated the Black community to wise up to the reality that the master's tools cannot be employed to dismantle the master's house, scholars of the Black diaspora took note. That statement resonated deeply with each and every student of diaspora studies then and now. But while the statement prompted many to action, thoughts about what tools would effectively tear down the walls of racism in an era where white supremacy is/was running rampant, were hard to envisage. The real concern, though, was how to dismantle an institutionalized system that is so entrenched that the odds for success are quite low. Along with Audre Lorde's (2018) challenge came the psychological clarity, grounded in Maslow's hierarchy of needs theory (1943), in which it is implied that until humans satisfy basic needs for survival, it is improbable that they would construct effective action for dismantling the systems of oppression that render them powerless to act strategically.

Climate change has risen to the apex of the developmental landscape. The industrialized world has made it nearly impossible for regions of the African diaspora to survive in a world free from cultural, economic, environmental, political, and social disaster. As the world watches the rise in sea levels and the impact of escalating temperatures, generating wildfires in some regions and devastating storms in others, there is no question that developing nations, particularly those of the African diaspora in the Americas and the Caribbean, are losing hope of finding affordable solutions to problems they have arguably not created for themselves. Recognizing that their reliance on industrialized partners over the decades of their political independence has embroiled them in partnerships that have been extractive and exploitative, the time has now come for them to take urgent action to mitigate the harm to development that climate change threatens.

Regenerative strategies have the capacity to shift attitudes away from the extractive behaviors of the past. Debates surrounding how climate change should be addressed, while being general to the international community, have less diffuse and more specific meaning for the diaspora, which can ill afford to address the abstract arguments of how much fossil fuel should be curtailed in managing climate change rather than how to narrow the gap in development between perpetrators of carbon spread and victims of that spread. It is in this context that our manuscript has attempted to offer a framework for mitigating climate effects and assuring sustainable development. As Figure 1.1 shows in this book, knowledge capital when identified in diasporic and other developing communities, may be produced at the community level to forge local and regional development and lead to the achievement of sustainable goals. The connection between the erosions from continued disaster and the attainment of sustainable development is not lost on the diaspora. For too long the complexity of inequalities separating wealthy from poor economies and the capacity of those able to address solutions versus those who are disproportionately victimized by inequities have led to staggering inefficiency, crises in leadership, and loss of resilience in seeking innovative strategies for sustainability, across the diaspora.

As observed in chapters 2 through 5, collaborative energies at the grassroots level, building on traditional, homebased practices and supported by communities empowered by their can-do spirit can generate new ways to thrive. Knowledge economies have always had the potential to mitigate loss. Indeed, many Western economists, including those leery of the concept of *knowledge* as an economic term, join scholars of philosophy, management, politics, and more in recognizing that knowledge economies may be the next stage in global economic development, succeeding the agricultural age, with its focus on land, followed by the industrial age with a focus on capital and labor (Weber 2011). Although critics of the term *knowledge economy* dismiss it as rhetoric in conversations related to globalization, there seems to be adequate recognition that regions that cannot engage technology in an advanced and meaningful way to maximize profits and claim development should not be afraid to find other, unique solutions to development.

Our work acknowledges that given the innate tendency to resiliency among people of the African diaspora, choosing a conceptual framework for identifying and producing the knowledge capital reposing in Afro-diasporic communities, can lead to innovative ways to sustain community development. In this context, our work began with an evaluation of the literature on equity and democracy, where expectations arising from this form of political leadership are exposed. Democracy assumes fair access to rights and freedoms, fair distribution of economic resources, and human well-being. Where promises of democracy fail, distrust in the system creeps in and populations may seek recourse in authoritarian promises for a better life. Such a life requires citizen responsiveness and policy implementation of the goals demanded by the populace. Disaster-prone communities prefer leadership before a catastrophe occurs or recurs. In collaboration with relevant stakeholders, they advocate for increased opportunities to raise their voice and propose action, in the pre-disaster stage, for how they might best mitigate the threats of impending disaster as climate vulnerabilities loom large.

First, the need to determine areas of vulnerability may necessarily vary from one diasporic locale to the next. While the need for ethnic equity and environmental justice or resource management may be strong in some communities within the three regions under study, energy concern and trauma recovery may be predominant in other regions. Understanding that an evaluation of the past can help shape strategies for the future by creative development of solutions, it is in communities' best interest to scrutinize those lessons of the past as they specifically apply to the vulnerable area to ensure that the challenge likely to be posed by the next catastrophe is met with successful planning and implementation of innovative solutions.

Each of the case studies of the three regions under scrutiny in this volume illustrates how post-disaster communities have learned from the past and how they have differentially developed innovative strategies to overcome the negative impacts of sociopolitical and environmental disaster in order to mitigate the crippling effects of climate change and its resultant stagnation of development. Any progress in disaster planning and successful strategizing made in these three regions may be emulated to scale by other members of the developing world, especially those of the African diaspora, whose territories have the potential to adapt meaningful practices to their present needs and empower communities to engage in knowledge sharing for the benefit of their regions.

Fundamentals of Climate Justice

The assumption that the tools for dismantling injustice and attaining self-determination are more specific than abstract, requires some conversation on what a restoration of climate justice, economic security, and gender equality would contribute to the dismantling of the master's house. Such a conversation would entail in diasporic societies a listing of what is included in a consideration of this theme. Climate justice requires:

1. An identification of community knowledge on what communities experienced before the climate warmed, what made that experience acceptable, and what needs to be done to restore to communities the sense of well-being residents felt when the air was clean and breathable, the water free from lead and other toxins, the soil not contaminated by poor sanitation (rendering roots and legumes wholesome and healthy), the rivers, lakes and streams unpolluted (creating no danger to marine life), and the seafood edible with no chance of mercury ingestion.
2. An understanding of what allows residents to thrive and feel economically secure.
3. An appreciation of community and the resilience with which communities collaborated to achieve a good life.
4. A recognition of the fact that without a holistic approach, incorporating the knowledge capital of all stakeholders—women, men, technocrats, business partners, cultural leaders, and others—tackling the issues of disaster preparation and mitigation cannot be successfully accomplished.

These parameters provide the foundations for communities to take steps to ensure residents enjoy the quality of life and food sources they once had. Some of the strategies engaged in Louisiana's communities or in Haiti's or Rwanda's may serve as a template. Restorative action may require new ways of participating in the life of the community, be it through alliance with nonprofits, agronomists, economists, fisher folk, food producers, political action committees, or others. Working with partners and allies in the struggle to claim one's best life is what diasporic communities have done through the ages. And, as elders have stressed in small communities, "experience teacheth fools." Learning from elders and sharing their experiences is a practical and successful way of retrieving cultural data that diasporic communities may have lost with gentrification and eagerness to keep up with other more modernized communities.

Economic Security

Economic development is perhaps the most difficult challenge diasporic communities have experienced. It is in itself a disaster from which few vulnerable communities recover. How do communities figure out what sustainable development even means in the absence of environmental security? There is, also, no easy way to grasp the concept of development without understanding that successful economic development efforts require innovative planning and a strong solutions-based approach to ensuring that the community is a comfortable, engaged, thriving place in which to live. Only then can a knowledge economy grow from the ground up to harness the knowledge capital resident in the community and be ready for production of holistic planning that includes short- and long-term insight in sync with community needs. Knowing the financial state of the community, the intellectual resources within it, and the strategic vision that will help to realize incentives for growth is an undeniable basis for development. Who is involved in the planning, the intellectual capital they bring to bear on strategic planning and incentivization, the project structures with potential for growth, and the related services and resources needed to implement the plans agreed—all are integral to the success of the initiatives to be undertaken for sustainable development (Akoh et al. 2011).

Communities in the Black diaspora are conscious of the trade-offs that must be made to ensure the benefits foreseen. Financial planning, bargaining capacity, fiscal allies, technological expertise, and the resilience of stakeholders can together create change and experience regional economic recovery and growth. While this is true for most communities, the exclusion of diasporic communities from planning centers, financial institutions, economic literacy, and a history of non-collaboration born of distrust (colonial powers ensured that would be the case), requires the diaspora to reflect on the obstacles to equity of the past and forge ahead with agency and voice, ensuring that their story is heard and their needs understood. As is recognized, doing nothing to change one's situation does not yield growth. Or, as reported, Albert Einstein wittily claimed, "Insanity is doing the same thing over and over and expecting different results" (Wilczek 2015).

Thus, at the heart of this volume is the recognition that development cannot take place without incremental economic security, which ultimately leads to sustainability. Yet, this is not a philosophy that resonates easily with diasporic communities where residents have suffered poverty and the ravages of the climate in silence waiting for the prosperity that remains untapped and hoping that by being resilient, tolerating their condition, a breakthrough might someday arise, given their intellectual capacity to thrive. Guyana was just beginning to recognize that their geological discoveries meant a turn in their fortunes when they were made brutally aware of the conflict between production of fossil fuels in an era of climate change and the potential harm it could hold for the country's economic advancement. How does a developing country in the diaspora respond to the knowledge that, like the rest of the world, it must limit a reliance on fossil fuels to mitigate the risks of climate change just when it is within reach of the very product that promises its escape from economic insecurity?

Gender Equality

There is no gainsaying the requirement for gender equity in the management of disaster. The knowledge production model proposed in the conceptual framework underpinning this study highlights the importance of empowering communities to capitalize on knowledge resources. This simply means ensuring that legitimate and full access to information, education, and planning be available to the whole community and that the latter be empowered if knowledge is to be effectively and successfully produced. Data have shown that a breakdown of the barriers to education and training for women has increased the potential of growth in the marketplace of ideas and planning. The gender equity index (GEI), as disclosed in the European Institute for Gender Equality (2022), supports the focus on women's empowerment and companies' commitment to gender equality in the workplace. Societies recognize that human life cannot be improved if only a fraction of humans are included in a culture of development. Diversity, inclusion, and excellence are the slogans that successful enterprises are using to proclaim their commitment to attracting and retaining the top talent in industry. Advocacy for women's voices to be heard and women's talents to be included in development plans strongly resonate within the international community and across international organizations, where resolutions embedded in the Convention on the Elimination of All Forms of Discrimination Against Women (CEDAW) form the only human rights treaty that addresses the rights of women (UN OHCHR 1981).

Although a majority of countries in the diaspora and most of the developed world have signed and ratified the Convention, the United States Senate has shied away from ratification, causing global concern for women's rights in the United States but more importantly for the lack of support of women globally in their fight against discrimination. From their perspective, US policymakers see ratification as having the capacity to potentially undermine US sovereignty and impact the private conduct of US citizens. Ratification of the CEDAW Convention remains a contentious issue in the chambers of Congress but is fully supported at the grassroots level where the Me Too movement continues to advocate strongly for the rights of women to self-determination and its resistance to organizations that seek to oppress women's right to personal freedoms, including a claim to reproductive justice. Black and other women of color are all too familiar with the role they play in advancing society, and recognize that, against all odds, they must be included in strategic planning for the environmental health of their communities.

Summary of Lessons Learned

Louisiana

A summary of the findings in chapter 2 reveals that Louisiana communities have benefited much from the collaborative energies they have gained by working with key elements of the Gulf Coast regions where concern for development and the urgency of overcoming obstacles posed to development by climate change and the existential crisis that attends it is paramount. For the most part, activism in Louisiana communities has been spearheaded by the Gulf Coast Center for Law and Policy (GCCLP). The strength of this organization has been seen in its outreach to neighboring frontline states along the Gulf Coast with strong thematic focus on democracy (equity, policy leadership); capacity management (knowledge and natural resources); advocacy (building community resilience, finding home-relevant solutions); and transitions (training for workforce shift into a new economy). In addition, Louisiana's internal outreach rested in its collaborative energies. The figure below illustrating the pillars on which Louisiana's energies are reposed may be a useful template for diasporic communities still unsure of how to build a strategic plan for mitigating climate risk, of which infrastructural change is a high priority. In this context, the work of grassroots community coalitions within the GCCLP, the Gulf of Mexico Alliance (GOMA), The Water Collaborative of Louisiana (TWC), the RISE movement in St. James Parish, the Concerned Citizens movement in St. John Parish, policymakers at City Hall and on the City Council of New Orleans, the Deep South Center for Environmental Justice (DSCEJ), the Walls Project (Baton Rouge), and other community organizations within and across Louisiana have all played a key role in dismantling forces of environmental oppression and enlightening communities on pathways to resist and disrupt the status quo. Although the chapter on Louisiana has done much to share examples of how local knowledge capital has been produced and converted into creative and innovative patterns of development, much of the supportive work done within rural and remote areas of the state has not been explored. This is one of the reasons why this volume may be helpful in understanding catastrophes from different contextual perspectives. Knowledge sharing is critical to the expansion of new intellectual horizons from which an engaged diaspora may build on the successes of other diasporic communities in the developed and developing world. Such knowledge sharing assumes a broad range of practices, including Town Hall meetings, creative arts exposés, and a myriad of community activities planned by groups whose heritage is retold in the story of Mardi Gras Indians, the Baby Dolls, food, music, and much more.

Figure 7.1 Louisiana. Pillars of Collaborative Intellectual Energy. The Water Collaborative (TWC). Developed by Jessica Dandridge.

The Walls Project, administered in Louisiana's capital city, Baton Rouge, is, for example, one of the creative educational solutions that highlight the innovative and indigenous, artistic spirit that communities in the capital city are now proud to share.

Haiti

Recent reports on Haiti's disaster experience have identified colonial exploitation as among the most egregious causes of its underdevelopment. A *New York Times* (NYT) investigation (Gamio et al. 2022) revealed that Haitian payments to France following that country's independence in 1804 has cost it $21 billion dollars and crippled its ability to achieve sustainable development. NYT calculated that Haiti's economy would have gained at least $21 billion over time had France not extracted $560 million payable to French banks and former "enslavers" after the country ended slavery and seized independence more than two centuries ago. Not only was Haiti initially denied the right to self-determination, but it was forced to pay for claiming that right. Persecution of Haitian leadership by colonial perpetrators and distrust of leadership by Haitian citizens over the years have deprived Haiti of the community it needs to enjoy a sustainable environment in which to grow. Climate change and its damage to Haitian territory, its economic security, and its national cohesion have exacerbated the crises Haiti now faces.

Haiti's catastrophic experiences may provide new insight into an understanding of knowledge capital identification and sharing, and its impact within the diaspora. So often we assume that people learn only from examples of what to do and how to develop. Haiti holds many examples of what not to do while still demonstrating also how resistance, though often problematic and accompanied by new hardships, may advocate for and empower those in the diaspora who are identifying creative and innovative ways toward recovery and development. Professor Wesley Alcenat, a Haitian American historian, has pointed out both in his public and academic scholarship that although Haitians are often seen as unknowing of and underdeveloped by the West, they are indeed better aware of the source of their exploitation than the French and the American public appears to be (Democracy Now! 2022). Similarly, the NYT and Haitian scholars call attention to the maltreatment Haitians have received by US authorities when fleeing post-disaster conditions in the region following the 2010 earthquake and later. As Alcenat argued, Haiti serves as a metaphor for understanding the colonial legacy of racism in the United States and France and their preference for other groups. Haiti's fate is connected with the engineering of the world toward imperialism (see Democracy Now! 2022; Horne 2015, 2020).

Certainly, in the area of how media can connect and inspire communities as well as how matters of governance may be curated and redesigned to yield the kind of national unity that Haiti lacks, there is much to learn and share. When disasters and other catastrophes add to the daily cares of small diasporic communities and even the past does not yield lessons for restorative healing and community empowerment, struggling economies such as Haiti's do not envision hope for recovery and sustainable development at the hands of Western governments but may certainly enlighten other Afro-diasporic communities on the realities of their struggle while learning from them of mechanisms states like Louisiana, equally maltreated by the French, have set in place to restore an infrastructure better poised to manage the risks of climate change.

The Haitian chapter evaluates the potential input of social networks to enliven communication and connectivity among Haitians while empowering Haitians at home and abroad to revive a homeland that has been buried under crisis for too long and needs to reclaim a heritage of resilience that may impatiently be seeking an opportunity to resurface and blossom into the economic, political, and sociocultural turnaround Haiti desperately needs. During Black History Month, a book produced by Dr. Sujan Kumar Dass (2020) proclaimed in its title that "Black People Invented Everything." The book's subtitle offered to share the deep history of indigenous creativity. Although a study in similar vein, "When the World Was Black," was published seven years earlier (Bailey 2013), very few readers in Africa and its diaspora are aware of the existence of such texts, perhaps because, unlike the spread of stories of diasporic peoples in the United States for whom ethnicity and skin color are such political weapons, historical narratives on Blackness and resilience have to be recounted whenever possible. A recognition of Black History during Black History Month helps engage the educational curriculum on diaspora studies in the United States but is rarely extended to the vast African diaspora.

Without deliberative knowledge sharing for the purpose of enlightening the Black diaspora and engineering opportunities for rebuilding infrastructure, economic growth, and development, many in the Americas and the Caribbean will fail to receive the empowerment and advocacy offered in knowledge economic narratives that aim to inspire and restore, through knowledge of the past, successes in the present, and potential for the future. Haiti's recovery can be as restorative as Rwanda's. Through educational and cultural portals, Haiti too can reclaim its sense of self and restore its people to the dignity and integrity it once enjoyed within and beyond the Black diaspora. Coalitions of Haitians both within the disaster-prone state as well as across its own scattered diaspora, via its creative arts initiatives and advocacy of local initiatives for development, need to empower each other with the resilience of their ancestors, and work assiduously toward achieving environmental and developmental sustainability. Despite those who suggest that Haiti is doomed by its "progress resistant genes" (Brooks 2010), finding new ways to connect with its past for the betterment of its future is viable within a knowledge economic model of development.

Rwanda

Our work on Rwanda has richly clarified the practicality of returning to the past to change the future. Although there is literature arguing against colonial interference as the cause of the genocide in 1994, recent scholarship (see Heldring 2020) asserts that the cause of the genocide reflects a colonial legacy of division, which was mitigated by pre-colonial institutions and arrangements. There is merit to this assertion, given the many ways European colonial leadership sought to exacerbate long-standing hostility between ethnic groups for their own exploitative benefit. It is also remarkable, when one observes that a return to pre-colonial institutions and arrangements to settle the injustices perpetrated during the genocide, now contributes to Rwanda's post-genocide recovery. We know that a preference for traditional courts, known as Gacaca, and sentencing in the restoration of justice to Rwanda since 1994 has been responsible, in large part, for the recovery now hailed by global onlookers as transformative for Rwanda. If indeed there were institutional arrangements in pre-colonial days that were used to help Rwanda recover in the twentieth century and onward, then there is clear justification for a call to return to and learn from historical roots rather than simply adopting the premise of the Western world that diasporic countries should follow the path and preconditions for development laid down by the West.

Knowledge must be shared for development to be sustained. Rwanda's experience may not be Jamaica's or Barbados's but there are elements of it to be seen in Guyana and some areas of Trinidad and Tobago where ethnic rivalries retain the potential for ethnic clashes between the early trans-shipment of African slaves and the later importation of indentured Indian labor to meet the economic needs of the European colonial offices. The same may be said for clashes among diasporic members of Latin American communities where indigenous populations, Afro-Latine, and Garifuna communities have engaged in tussles for economic survival. Until fragmented groups accept the old proverb that "if you want to go quickly, go alone, but if you want to go far, go together," the realization of collective nation-building may remain dormant. Speed and distance are morally neutral objectives. People must understand what it is they wish to achieve and how that might best be achieved. Pondering these old African proverbs can be empowering for diasporic communities. This particular proverb is easily applied to a number of issues facing communities; issues of engaging women and men in collaborative development for national growth despite the ancient practice of designing roles for women that are different for men, or dismissing the knowledge capital that comes from one group or another as deficient, thus failing to maximize the benefits of holistic planning and achievable development. Sustainable communities of the African diaspora stand to gain much from reflecting on the wisdoms of the past and engaging them in such ways as to realize development for posterity.

Other Areas of the Black Diaspora

In the Caribbean, climate change policy, representing the will of the Caribbean people and their representative leaders, has had a noticeable impact. In a project report submitted to the Commonwealth Secretariat, it is noted that Jamaica established a Ministry of Water, Land, Environment and Climate Change (2012) and launched, under its umbrella, a Climate Change advisory Committee (CCaC) with a membership of academia, the private sector, government agencies, civil society, and environmental nongovernmental organizations in 2013. CCaC has been charged with the responsibility of giving strategic direction to the ministry, the Cabinet, and the country as a whole. The committee will address all matters related to climate change, the development of a Climate Change Policy, a Climate Change Department, and guidance for the implementation of public awareness and public education programs (Pickersgill 2013).

The Sir Arthur Lewis Institute of Social and Economic Studies in Jamaica has taken the lead in building the research effort in the Caribbean region to make Caribbean Communities climate-change resilient. The immediate aim is to engage all fifteen members of the community and its five associate members so that they become more aware of and educated on matters related to climate change, to influence their actions. In like manner, other communities in the diaspora must engage in activities with national and international partners to build resilience collectively in order to mitigate and adapt to climate change.

In Guyana, something of a paradox exists. While communities there recognize that reliance on fossil fuels must be speedily reduced in the face of climate effects such as desertification, soil erosion, deforestation, and the extinction of unique species, they are also aware, based on IMF data projecting huge growth in GDP (IMF 2019), that they cannot ignore the fact that the projected windfall is from extractive sources. The dilemma for local planners, then, is how to increase economic development while decreasing reliance on fossil fuel extraction, when the latter provides their only opportunity for closing the economic gap of the past fifty-seven years since gaining independence from Britain. In addition, though learning from the past usually provides helpful road signs on what to do or not do in the present, Guyana's history with natural resource extraction and her tendency, in volatile political situations, to abandon one development plan for another without implementing any or documenting lessons learned, is both a scary and an exciting prospect (David 2021). Thus, recommending a knowledge economic model for Guyana is not without its dangers. Communities and individuals outside of the government arena have little motivation to explore their knowledge capacities and even less for embarking on innovative and creative designs for development.

It is precisely why the Guyanese team of young Webby Award winners caught the attention of so many well-wishers (Callender 2022). To creatively pitch a design that would be useful to communities in their effort to mitigate the risks of climate change creates an opportunity not only for conservation-minded Guyanese but also for the remaining diaspora, some included among the other 150 countries keen enough to save the planet, as to participate in the 5G for Change Hackathon, held in New York in 2022. Innovation within the developing world and, primarily, the diaspora can become contagious and, with knowledge sharing, effective designs for combating risk may facilitate development and aid in disaster preparedness.

Conclusion

Our manuscript set out to tackle a number of interrelated questions pertinent to the lives of African-descended people, many of whom were initially trans-shipped to the "new world" and have now settled in many regions of the world. It is believed that the largest concentration of people of African descent is estimated at approximately 150 million and reside in Latin America and the Caribbean. In addition to their displacement from Africa during slavery, many have migrated to Europe, Asia, and also within Africa, seeking a better life away from continual marginalization and the discrimination they suffer as the historic legacy of the slave trade. The search for development for descendants of Africans has been marked by centuries of crisis, ranging from environmental to political to cultural to social but most markedly to economic catastrophe. Thus, the task for this book was to craft an intersectional and interdisciplinary study of how the Black diaspora might reclaim its innate talents to find alternative ways to reboot their energies and forge a trajectory of growth that may change the flattened or downward spiral that has characterized progress toward an upward linear curve that symbolizes hope for the future.

The conceptual framework envisioned for this study allowed authors to look at the margins of development to address impacts of the social, political, economic, and cultural structures that have inhibited the development of people throughout the African diaspora and stagnated the advancement of their self-determination in societies, traumatized by barriers to, and inequality in, their access to basic human rights in areas such as quality education, affordable housing, climate justice, and viable healthcare. Lacking these basic human necessities for survival and living in geographical spaces that further exacerbate the hazards confronting African descendants, a new model for development seemed necessary in the era of climate change and the existential threats to the humanity of diasporic peoples. This is not to say that we do not recognize the suffering of those in developing countries all over the world. Instead, we intentionally focus on the Black diaspora because this group of humans is consistently found at the bottom of all measures of growth the world over. Without inclusion of all those whose vulnerabilities never get addressed, change will never be experienced by those at risk of being left behind, forgotten, broken, and without hope of ever achieving the good life other humans are promised and enjoy.

In a recent announcement of a creative project on tech innovation, the innovation of an artificial intelligence sound monitoring device program to tackle the issue of illegal logging in Guyana (The Webby Awards 2022) was awarded to four Guyanese students, alumni of the University of Guyana. This team of innovators pitched an argument described as innovative and practical to win the prestigious inaugural award of the 5G for Change Hackathon. Their design, to tackle deforestation, was hailed by judges to be the best submission from teams in over 150 countries at the 26th Annual Webby Awards, held in New York on 17 May. As an innovation with impact for climate change, this award has relevance not only for the diaspora but for other countries facing issues of development that need to adapt to novel homegrown methodologies to address climate risks such as deforestation, coastal erosion, and economic insecurities that contribute to further damage of the ecosystem. The award winners explained (see Callender 2022) that since over 80 percent of Guyana's land mass is forested, illegal logging negatively impacts the country's forestry and is responsible for 30 percent of carbon emissions that come from clearing the area. The device they created will use sound monitoring and drones to mitigate illegal logging and the ease with which violators "bypass government regulations through bribery" (The Webby Awards 2022). Creative innovations such as these validate a belief that innovative information communication technology created at the grassroots level and shared with Black diasporic communities can play an integral role in empowering emerging societies not only to learn more about climate change issues such as deforestation, propagation of fossil fuels, carbon spread, and its devastation but also to advocate for greater knowledge sharing and feedback toward the generation of new productive knowledge capital.

Because this volume aims to engage work on disasters in order to influence policy formation and implementation where necessary, it seemed appropriate to take a broad look at catastrophe as inclusive of environmental, human, and political trauma of the past, present, and threatening to the future. Identifying localities, in both the developed and developing world, experiencing environmental hazard and injustice, economic insecurity and unequal access to labor markets, and barriers to nation-building in the form of the devaluation and exclusion of women's capacities to contribute holistically to policy-making for societal development, this study has made a valiant effort to connect the dots intertwining development, in the hope that readers may gain new insights into another, more self-reliant model, for sustainable growth.

Although we are aware that the diaspora is unique in its whole, we recognize that it is quite diverse in its needs and must tailor development to a scale of societal need. To serve as exemplars of diverse need, three states were selected—one from the African continent (Rwanda), one from the developing Caribbean world (Haiti), and one from the developed North (Louisiana)—whose experiences with disaster and innovative approaches to managing environmental injustice, economic insecurity, poverty, and social capital inclusive of gender constraints, may serve either as a blueprint for others in the diaspora or as a tool of empowerment for and advocacy of knowledge production to domestic scale. At the heart of our study is the belief that justice averted or postponed is justice denied. The diaspora has struggled under the weight of racism and discrimination for so long that many have abandoned pursuit of a better life with self-determination and wait to achieve their just desserts in an afterlife that their faiths have led them to aspire to with unquestioning devotion. Those of us who believe in personal freedom and full self-determination understand that, adapted from the words of Stephen Grellet (1773–1855), "I expect to pass through this world but once; any good thing therefore that I can do, or any kindness that I can show to any fellow-creature, let me do it now; let me not defer or neglect it, for I shall not pass this way again" (Ratcliffe 2016).

In the beginning of this study, we emphasized that our focus would be unlike other authors' foci on disaster-related sustainable development, who have primarily centered their writing on top-down solutions. Both the academic and professional exercise of managing disaster crises have only looked to governmental and nongovernmental models for disaster recovery. This volume has opted to provide an alternative prescription that centers "bottom-up" models that foreground the knowledge and innovations that are native to the places where the catastrophe strikes. We have therefore presented a theoretical framework of knowledge economies in post-disaster societies that has been laid out through a trans-diasporic analysis of crises in select states and localities that are part of the Black diaspora. Our work has been guided by Black feminist scholarship, such as Audre Lorde's, who has argued that we cannot use the "master's tools to dismantle the master's house." We have thus interrogated meaningful ways to progress, identifying the failure of exclusively foreign-based, tech-based solutions prescribed by the Global North to have continuously dismissed the Global South and Black localities in the Global North, and proposed new pathways via knowledge production as a supplemental method for exploring community-based, post-disaster sustainability.

Through a multidisciplinary and interdisciplinary review of the production of sustainable knowledge economies from the ground up, including scholarship in a cross section of fields, the editor and contributors to this volume have hopefully re-centered and re-framed how we approach, design, and execute sustainable development in post-disaster societies. We hope to share the knowledge found within post-disaster societies to provide a blueprint for a viable and equitable recovery in vulnerable, diasporic spaces that have experienced both natural and man-made crises. Hopefully, we have met our goal to advocate for racial, restorative justice through sustainable development as societies across the diaspora continue to disproportionately recover from crises. Hopefully, too, we have been clear in our belief that the cultural wisdoms of the past will lead the Black diaspora to reclaim the resilience of its ancestors and rebuild, with communal strength, the sustainable future to which African-descended peoples are entitled.

Pamela Waldron-Moore, the Leslie R. Jacobs Professor at Xavier University of Louisiana, holds a PhD in political science with specialization in Comparative Politics and International Relations. She has taught at universities in the Caribbean and the United States. Among her research interests are the political economy of development, gendered disparity, and climate justice. She is published in peer-reviewed journals of comparative politics and international studies and contributes regularly to scholarly forums on transformative pedagogy. She is a recipient of Keynote Speaker awards for research on women and the 2018 Jewel Prestage Mentorship Award from the National Conference of Black Political Scientists (NCOBPS).

References

Akoh, Ben, Livia Bizikova, Jo-Ellen Parry, Heather Creech, Julie Karami, Daniella Echeverria, Anne Hammill, and Philip Gass. 2011. "Africa Transformation-Ready: The Strategic Application of Information and Communication Technologies to Climate Change Adaptation in Africa." International Institute for Sustainable Development, 5 March. Retrieved 23 April 2023 from https://www.iisd.org/publications/report/africa-transformation-ready-strategic-application-information-and-communication.

Bailey, Robert. 2013. *When the World Was Black: The Untold History of the World's First Civilizations*. Atlanta, GA: Supreme Design Publishing.

Brooks, David. 2010. "Opinion: Did Haiti's Poverty and Culture Worsen the Tragedy." *New York Times*, 15 January. Retrieved 23 April 2023 from https://www.nytimes.com/2010/01/16/opinion/lweb16brooks.html.

Callender, Jessica. 2022. "Guyanese Team Wins First-Ever Webby Hackathon Award Ahead of Over 150 Countries." *mtvgy*, 17 May. Retrieved 23 April 2023 from https://www.mtvgy.com/post/guyanese-team-wins-first-ever-webby-hackathon-award-ahead-of-over-150-countries.

Dass, Sujan Kumar. 2020. *Black People Invented Everything: The Deep History of Indigenous Creativity*. Atlanta, GA: Supreme Design Publishing.

David, Shameza. 2021. "The Paradox of Being an Oil-Producing State and Pursuing a Green Development Agenda." *Academia Letters, Article 3878*. https://doi.org/10.20935/AL3878.

Democracy Now! 2022. "Debt, Coups, and Colonialism in Haiti: France and US Urged to Pay Reparations for Destroying Nation." YouTube, uploaded 24 May. Retrieved 23 April 2023 from https://www.youtube.com/watch?v=X2q98odZz2U.

Gamio, Lazaro, Constant Méheut, Catherine Porter, Selam Gebrekidan, Allison McCann, and Matt Apuzzo. 2022. "The Ransom: Haiti's Lost Billions." *The New York Times*, 20 May. Retrieved 23 April 2023 from https://www.nytimes.com/interactive/2022/05/20/world/americas/enslaved-haiti-debt-timeline.html.

Gender Equity Index. 2022. "Core Domains." European Institute for Gender Equality. Retrieved 23 April 2023 from https://eige.europa.eu/gender-equality-index/2022.

Heldring, Leander. 2020. "The Origins of Violence in Rwanda." *The Review of Economic Studies* 88(2): 730–63. https://doi.org/10.1093/restud/rdaa028.

Horne, Gerald. 2015. *Confronting Black Jacobins: The U.S., the Haitian Revolution, and the Origins of the Dominican Republic*. New York: NYU Press.

———. 2020. *The Dawning of the Apocalypse: The Roots of Slavery, White Supremacy, Settler Colonialism, and Capitalism in the Long Sixteenth Century*. New York: NYU Press.

International Monetary Fund (IMF). 2019. "IMF Staff Country Report: Guyana: 2019 Article IV Consultation-Press Release; Staff Report; and Statement by the Executive Director for Guyana." IMF, 17 September. Retrieved 23 April 2023 from https://www.imf.org/en/Publications/CR/Issues/2019/09/16/Guyana-2019-Article-IV-Consultation-Press-Release-Staff-Report-and-Statement-by-the-48678.

Lorde, Audre. 2018. *The Master's Tools Will Never Dismantle the Master's House*. London: Penguin Classics.

Maslow, Abraham H. 1943. "A Theory of Human Motivation." *Psychological Review* 50(4): 370–96.

Pickersgill, Hon. Robert. 2013. "Foreword." In *Small States Digest: Globalization, Climate Change and Rural Resilience in the Caribbean* 2, 4–5. n.p: Commonwealth Secretariat.

Ratcliffe, Susan, ed. 2016. "Stephen Grellet 1773–1855." In *Oxford Essential Quotations*, 4th edition. Oxford: Oxford University Press. Retrieved 23 April 2023 from https://www.oxfordreference.com/display/10.1093/acref/9780191826719.001.0001/q-oro-ed4-00005064?rskey=M4sQZo&result=1606.

The Webby Awards. 2022. "FramePerfect Tackles Illegal Logging in Guyana." *Webby Awards*. Retrieved 23 April 2023 from https://www.webbyawards.com/5g-for-change-finalist-team-frame-perfect/.

United Nations Human Rights Office of the High Commissioner (UN OHCHR). 1981. "Convention on the Elimination of All Forms of Discrimination against Women (CEDAW)." *United Nations*, 3 September. Retrieved 23 April 2023 from https://www.ohchr.org/en/instruments-mechanisms/instruments/convention-elimination-all-forms-discrimination-against-women.

Weber, Allan S. 2011. "The Role of Education in Knowledge Economies in Developing Countries." *Procedia: Social and Behavioral Sciences* 15: 2589–94

Wilczek, Frank. 2015. "Einstein's Parable of Quantum Insanity." *Quanta Magazine*, 10 September. Retrieved 24 April 2023 from https://www.quantamagazine.org/einsteins-parable-of-quantum-insanity-20150910.

Appendix I

Although knowledge economies are built on the production of knowledge through innovative ideas for sustainable development, little is known in the Black diaspora of the African American and Caribbean innovations that have contributed to global advancement. Without knowledge sharing, there is little incentive, in today's world, for knowledge economies to thrive. Below is a brief glimpse into inventions by Africans in the diaspora that may have gone unnoticed:

Overview of Famous Inventions and Their Inventors

Breedlove, Sarah, better known as Madam C. J. Walker (1867–1919): Created her own hair care system in the 1900s. She marketed the product via direct sales and donated much of her $600,000 wealth to the YMCA and NAACP.

Carver, George Washington (1864–1943): As an agricultural scientist and inventor, Carver promoted alternative crops to cotton and methods of preventing soil depletion.

Crum, George (1824–1914): Restaurateur known for serving thinly sliced fried potatoes, which led to the popularization of what are now known as potato chips in the 1860s.

Downing, Philip (1857–1934): Best known for two significant inventions: (1) the protective mailbox to keep mail safe from inclement weather and indoor safety, and (2) street railway switches.

Drew, Dr. Charles (1904–1950): A surgeon and medical researcher, Dr. Drew improved blood storage techniques and separated red cells from plasma to preserve blood for long periods of time.

Goode, Sarah (1855–1905): Invented the folding cabinet bed to fit in small homes in 1885. It is now known in modern-day as the Murphy bed.

Latimer, Lewis (1848–1928): Patented carbon filament for a light bulb in 1868 and assisted Thomas Edison in planning a light bulb model.

Lawson, Jerry (1940–2011): An African American electronics engineer, Lawson forever changed the video game industry by designing the Fairchild channel video game console and leading the team that pioneered the commercial video game cartridge. He has been dubbed the "father of the videogame cartridge."

Matzeliger, Jan (1852–1889): Born in Dutch Guiana (now named Suriname), he invented the automatic shoe lasting machine, mechanizing the process of joining a shoe sole to its upper. The invention revolutionized the shoe industry, producing 150–700 pairs of shoes per day in 1873, thus making shoe purchase more affordable.

McCoy, Elijah (1844–1929): Invented an automatic oil cup, in 1872, to lubricate the train's axel while it was in motion, so that fewer stops would be made en route. This improved safety and ensured timely arrival at destinations.

Morgan, Elijah (1877–1963): Developed the safety hood for firemen to survive smoke and inhale oxygen in 1916. He later created an early prototype of a three-position traffic signal.

Woods, Granville (1856–1910): Invented the multiplex telegraph, which allowed dispatchers and engineers to communicate in 1887. Among the fifty patents he held in the United States were the steam boiler furnace for trains and the development of streetcars.

Appendix II.

List of Known Black Inventors and Inventions.

Source: UCSD Library

INVENTION	INVENTOR
Airplane Propelling	James S. Adams
Air Conditioning Unit	Frederick M. Jones
Air Ship (Blimp)	J. F. Pickering
Alphabet	Africans
Arm for Record Player	Joseph Dickinson
Automatic Air Brake	Granville T. Woods
Automatic Fishing Reel	George Cook
Automatic Gear Shift	R. B. Spikes
Baby Buggy	W. H. Richardson
Bicycle Frame	Isaac R. Johnson
Biscuit Cutter	A. P. Ashbourne
Blood Plasma	Dr. Charles Drew
Car Coupler	Andrew J. Beard
Cellular Phone	Henry Sampson
Chess	Africans
Civilization	Africans
Clothes Dresser	John H. Jordan
Clothes Drier	G. T. Sampson
Coin Changer	James A. Bauer
Corn Planter	Henry Blair
Cotton Planter	Henry Blair
Curtain Rod	S. R. Scottron
Defroster	Frederick M. Jones
Disposal Syringe	Phil Brooks
Doorknob	O. Dorsey
Doorstop	O. Dorsey
Dustpan	I. P. Ray
Egg Beater	W. Johnson

Electric Cut-Off Switch	Granville T. Woods
Electric Lamp	Latimer and Nichols
Electric Mechanical Brake	Granville T. Woods
Electric Railway System	Granville T. Woods
Elevator	Alexander Miles
Envelope Seal	F. W. Leslie
Eye Protector	P. Johnson
Far-Ultraviolet Radian Detector	George Carruthers
Fire Escape Ladder	J. B. Winters
Fire Extinguisher	Tom J. Marshal
Folding Bed	L. C. Bailey
Folding Chair	Purdy/Sadgwar
Fountain Pen	W. B. Purvis
Furniture Caster	David A. Fisher
Galvanic Battery	Granville T. Woods
Gas Burner	B. F. Jackson
Gas Mask	Garrett Morgan
Golf Tee	George F. Grant
Guided Missile	Otis Boykins
Guitar	Robert Flemming, Jr.
Hairbrush	Lyda Newman
Hand Stamp	W. B. Purvis
Heating Furnace	Alice H. Parker
Helicopter	Paul E. Williams
Home Security System	Marie Brown
Horse Bridle Bit	L. F. Brown
Horse Riding Saddle	Wm. D. Davis
Horseshoe	Oscar E. Brown
Hydraulic Shock Absorber	Ralph Sanderson
Ice Cream	Augustus Jackson
Ice Cream Mold	A.L. Cralle
Image Converter	George Carruthers
Insect Destroyer Gun	A. C. Richardson
Internal Combustion Engine	Frederick M. Jones
Ironing Board	Sarah Boone
Kitchen Table	H. A. Jackson
Lantern	Michael Harney
Laser Fuels	Lester Lee
Lawn Mower	John A. Burr

Lawn Sprinkler	J. W. Smith
Letter Box	G. E. Becket
Lock	W. A. Martin
Lotions and Soaps	George Washington Carver
Lubricators	Elijah McCoy
Medicine	Africans
Mop	T. W. Stewart
Motor	J. Gregory
Multi-Stage Rocket	Adolph Shamms
Paints and Stains	George Washington Carver
Paper	Africans
Peanut Butter	George Washington Carver
Pencil Sharpener	John L. Love
Photo Embossing Machine	Clantonia J. Dorticus
Photo Print Wash	Clantonia J. Dorticus
Player Piano	Joseph Dickinson
Postal Letter Box	Phillip B. Downing
Potato Chips	George Crum
Pressing Comb	Walter Sammons
Pressure Cooker	Maurice W. Lee
Printing Press	W. A. Lavalette
Programmable Remote Controllers	Joseph N. Jackson
Refrigeration Controls	Frederick M. Jones
Refrigerator	J. Standard
Relay Instrument	Granville T. Woods
Rocket Catapult	Hugh Macdonald
Roller Coaster	Granville T. Woods
Rotary Engine	Andrew J. Beard
Sani-Phone	Jerry Johnson
Shoe	W. A. Deitz
Shoe Lasting Machine	Jan Matzeliger
Space Shuttle Retrieval Arm	W. M. Harwell
Stainless Steel Pads	Alfred Benjamin
Stairclimbing Wheelchair	Rufus J. Weaver
Starter Generator	Frederick M. Jones
Street Sweeper	C. B. Brooks
Sugar Refinement	N. Rillieux
Super Soaker	Lonnie Johnson
Telephone System	Granville T. Woods

Telephone Transmitter	Granville T. Woods
Thermo Hair Curlers	Solomon Harper
Toilet	T. Elkins
Torpedo Discharger	H. Bradberry
Traffic Signal	Garrett Morgan
Train Alarm	R. A. Butler
Two-Cycle Gas Engine	Frederick M. Jones
Typewriter	Burridge Marshman
Urinalysis Machine	Dewey Sanderson
Video Commander	Joseph N. Jackson
Window Cleaner	A. L. Lewis
Wrench	John A. Johnson

Index

adaptation, disaster recovery and, 5–6
Adaptation Fund, 41
Africa: gender equality in, 173–77; indigenous African culture and, 8–9; postcolonial sociopolitical perspectives in, 146–49
African Americans: as flood victims, 79–80; media exclusion of, 111–12; pollution and poverty for, 71–72
African-Caribbean-Pacific (ACP) partnership, 37
African diaspora. *See* Black diaspora
Akan culture, Sankofa and, 160–62, 182–85
Alcenat, Wesley, 202
Alterpresse (Haiti), 118
American Society of Landscape Architects, 87
Americas: climate change in, 194–96; gender equality in, 170–74; knowledge production in, 44, 172–74; post-disaster recovery and, 65–66; sustainable development in, 185–89; water management in, 179–81
AMIH (L'Association des Médias D'Haïti Indépendants), 116
Angelou, Maya, 190–91
ANMH (L'Association Nationale des Médias Haïtiens), 116
Aristotle, 32–33, 66
Army Corp of Engineers, 80–82, 86
Asia, development model in, 9–10, 34
Association of Haitian Journalists, 116
Atlantic-Live (*Atlantic Monthly*), 68
Aubert, Jean-Eric, 36

Barbados, decolonization of, 190
Baton Rouge, Louisiana: flooding in, 79–80; Walls Project in, 42–43
belling the cat (Caribbean fable), 164–66
Biden, Joe, 110
biologically-based community resilience, 137–38
Biracyaza, Emmanuel, 130–54
Black diaspora: climate change impact on, 2–3, 8–9, 21–22, 161–64, 194–96; cultural knowledge retention in, 39–41; disaster recovery and knowledge production by, 9–11; gender equality and, 170–77; history of, 7; information access and, 45–49; institutional economic structures of, 36–38; knowledge capital in, 7, 25–26; knowledge sharing by, 96–97; postcolonial sociopolitical perspectives of, 146–49; poverty and, 164–69; resistance to mental health interventions in, 138–39; Sankofa beliefs and, 5, 160–62, 184–85; scholarship on, 194; sites of investigation, 12–13; sustainable development and, 187–90
Black environmentalism, 72; grassroots coalitions and, 91–97
Black History Month, 202–3
Black Lives Matter, 14
Bonilla, Yarimar, 40–41
Bowie, Laura, 173
"Breaking the Thermometer to Hide the Fever" (McCalla performance), 186
Brookings Institute, 45
Brown v. Board of Education, 48

Bucket Brigade (Louisiana activities), 41–43
Burdick, David, 180
burial rites, in post-genocide Rwanda, 152–53
Burundi, postcolonial civil conflict in, 147
Butaro Hospital (Rwanda), 167–68

calpysos, post-disaster recovery and, 44
Campos, Karina, 57, 173
Cancer Alley (US Gulf region), 71–72, 90, 92–93
Caribbean Economic Community (CARICOM), 37–38, 171–72
Caribbean fables, Black diaspora and role of, 164–66
Caribbean region: climate change in, 194–96, 204–6; gender equality in, 170–74; knowledge production in, 44, 172–74, 204; post-disaster recovery and, 65–66; sustainable development in, 185–89; water management in, 179–81
Castle sisters, 187
Catch Basin (digital newsletter), 65–66
celebratory practices, post-disaster recovery and, 44
censorship, information access and, 47–48
Charles, Eugenia, 171
Chase, Dooky, 187
Chitimacha Tribe (Ojibwe), 73–75
chloroprene, health risks of, 92
Churchill, Winston, 140
civil society, Haitian social capital and, 107–8
climate change: Black diaspora and, 2–3, 8–9, 21–22, 161–64, 194–96; costs, benefits and outcomes of, 182–83; cultural knowledge systems and management of, 40–41; global collaboration on, 159–60; in Louisiana, 65–66, 81–82; post-disaster development and, 28–32; public health and, 178–81; sustainable development goals and, 181–84
Climate Change advisory Committee (CCaC) (Jamaica), 205
climate disaster, research on, 2–6
climate justice, 196–97
Coalition Against Death Alley (CADA), 93
Code of Ethics of the Haitian Press, 116
collective trauma: Louisiana experience of, 68–69, 83–84, 88–90; in post-genocide Rwanda, 152–53; Rwanda community resilience model and, 135–38
colonialism: African economics and, 25; disaster in Louisiana and, 69–72; education and training under, 38–39; in Haiti, 201–3; indigenous African culture and, 7–9; in Mississippi River Valley, 74–75; postcolonial sociopolitical perspectives on, 146–49; Rwandan genocide and, 131–33
command economics, Asian development model and, 9–10
Commonwealth Secretariat, 204–5
community collaboration: development and, 3–4; innovative practices and, 56; institutional economic structures and, 37–38; in Louisiana, 18–19, 188–89; Louisiana Bucket Brigade, 41–43; map of Louisiana communities, 97–98; natural infrastructure movement and, 89–90; Ndi Umunyarwanda policy and, 148–49; post-disaster recovery and, 29, 65–66; post-genocide recovery and, 142–50; Rwanda post-disaster recovery and, 4–5; seismic resilience and, 83–84
community development centers (Rwanda), 43
community justice in Rwanda, 133–35
community resilience model (CRM) (Rwanda), 135–38
Concerned Citizens of St. John, 92–93, 200
Condé, Maryse, 185
Coolidge, Calvin, 80
COP26 (climate conference) (Glasgow, 2021), 178, 187

Copernicus, 153
copyright laws, information access and, 47–48
cost sharing, sustainable development and, 56
COVID-19 pandemic: Black diaspora and, 21–22; in Haiti, 107; in Rwanda, 130–31; self-help and resilience in face of, 35–36
credibility, post-disaster Haiti and establishment of, 115–16
Crenshaw Williams, Kimberlé, 165
cultural knowledge systems: African diaspora and, 7–9, 31–32; community justice in Rwanda and, 133–35; gender equality and, 169; in Louisiana, 67–68; post-disaster development and, 13–14; in post-genocide Rwanda, 142; retention of, 39–41; social capital in Haiti and, 109–10; sustainable development and, 187–89
Cutteridge, J. O., 44

Dahl, Robert, 32, 46, 67, 92
dance: post-disaster recovery and, 44; in post-genocide Rwanda, 151–52
Dandridge, Jessica, 65–99
"Dan is the Man" (Mighty Sparrow song), 44
Danticat, Edwidge, 185
Dass, Sujan Kumar, 203
Death Alley. *See* Cancer Alley (US Gulf region)
decolonization, sustainable development and, 190
Deep South Center for Environmental Justice (DSCEJ), 200
demand-side economic theory, 30
democracy: equity and, 15–16, 29–30; gender equality and, 169; knowledge democratization in Haiti and, 112–14; sustainable development and limits of, 30–36
Denka Corporation, 92
development: challenges to, 49–55; democracy and, 32–36; globalization framework for, 13–14; indigenous knowledge and, 8–9; information access and, 46–49; Rwandan post-genocide resiliency *vs.*, 153–54; sustainability and, 3
dialogue national (national dialogue) initiative (Haiti), 124
digital technology: access to, 46–49; contemporary public sphere and, 110–12; Haitian digital networks and, 106–8; knowledge production and, 30; non-western structures for, 124–26; social capital mediation in Haiti and, 120–26
disaster recovery: adaptation and, 5–6; cultural knowledge systems and, 39–41; equity in, 15–16; gender equality and, 199; information access and, 46–49; knowledge democratization and, 112–14; knowledge production for, 2–3, 9–11; research on, 1–2, 28–29; sociocultural practices for, 130–54
distributive justice, 164–69
Dottin-Haley, Blair, 187
Dragozet, Aleksandra, 173
Du Bois, W. E. B., 13–14
Dupont Corporation, 92

earnings and income, gender inequality in, 170–72
earthquake of 2010, disaster in Haiti from, 29, 45, 105–8
economic development: economic security and, 197–98; Haitian earthquake impact on, 110; natural infrastructure movement and, 89–90; in post-genocide Rwanda, 130–31
economic liberalism: development linked to, 27–30; interventionist perspectives on, 15–16
education and training: in Caribbean cultures, 44; gender equality in, 172–73, 175–77; information access and, 47–49; knowledge economies and, 38–41; in post-genocide Rwanda, 134–35, 175

Ehrenwerth, Justin, 179–80
Emancipation Proclamation (1863), 76
employment, gender equality in, 171–72
environmental justice: fundamentals of, 196–97; income inequality and, 70–72; Indigenous peoples' chronicle of, 69–70; innovative systems for mediation of, 41–43
Environmental Protection Agency (EPA), 92
environmental scholarship, sustainable development and, 186
equitable development: democracy's efficacy and, 32–36; in disaster recovery, 15–16, 29–32
European Union-African Caribbean Pacific (EU-ACP) project, 170
Exxon Corporation, 168

faith-based communities, grassroots coalitions and, 91–97
family resiliency, post-genocide recovery and, 142–50
Federal Reserve policies, 30
FEMA (Flood Emergency Management Agency), 81
feminist politics, gender equality and, 170
5G for Change Hackathon, 206
Flood Control Act (1928), 80
flood management, sustainable development and, 79–82, 87–88, 179–81
Food and Agriculture Organization (FAO) (UN), 119
food deserts, 188
Foreign Affairs Newsletter, 12
foreign direct investment (FDI), disaster recovery and, 113
forgiveness, post-genocide recovery and, 142–50
Formosa Plastics, 91–92
"A Framework to Quantitatively Assess and Enhance the Seismic Resilience of Communities" (Bruneau), 83
France, Haitian payments to, 201–3
Francisco, Slinger, 44
Fuller, Lisa, 166
Gacaca community justice system (Rwanda), 133–35, 148–49, 204
Gallagher, Kelly, 12
gender equality: in Africa, 174–77; in Caribbean, 170–74; disaster management and, 199; earnings and income, 170–72; poverty and, 164–69; in Rwanda, 149–50, 174–77; sustainable development and, 20–21, 169–77
genocide: resilience and recovery from, 130–31; restorative justice and, 133–35; in Rwanda, 131–33; socio-therapy and recovery from, 138–41
Genocide Survivors Students Association (Association des Etudiants Et Éleves Rescapés Du Genocide, AERG), 136
Gentilly Resiliency District (New Orleans, LA), 87–88
Gerring, John, 33
gestures, community resilience model, 137–38
globalization: democracy and, 58–59; development policy based on, 13–14
Global North: educational structures of, 40–41; gender gap in, 171–72; limits of development solutions from, 26, 30–32; wealth distribution and, 163–64
Global South, poverty in, 164–69
government: distrust in, 33–34; excessive reliance on, 57; Haitian culture and, 110; Haitian demands for accountability of, 106–8, 119–20; Haitian protests and shutdown of, 123–24; responsiveness of, 46, 71–72; Rwanda post-genocide policies, 142–50
grassroots coalitions: Ayiti Kale Je (Haitian Grassroots Watch), 116–20; green infrastructure and, 90–97; Gulf Coast alliances, 93–95; parish coalitions, 91–93
gray infrastructure systems, failure of, 87–88

Greater New Orleans Urban Water Plan, 85–86
Great Flood of 1927, 79–80, 89
green infrastructure: development of, 87–90 (*See also* natural infrastructure movement); diasporic communities and, 9–10; grassroots coalitions and, 90–97; poverty reduction and, 167–69; in Rwanda, 189
Green New Deal, 94–95
green space, loss in New Orleans of, 81–82
grounding skills, community resilience model, 137–38
Group of 77, 37, 183–84
Gulf Coast alliances, 93–95
Gulf Coast Center for Law and Policy (GCCLP), 93–95, 187, 200
Gulf of Mexico Alliance (GOMA), 94–95, 200
Gulf South for a Green New Deal (GS4GND), 94–95, 187
Gurung, Jeanette, 173
Guyana: climate change policy in, 183–84, 205–7; economic development in, 198; election crisis in, 139; overthrow of colonialism in, 187; poverty reduction in, 168; water resource management in, 178–79
Güzel, Mehmet, 122–23

Habimama, Samuel, 43, 130–54
Haiti: Black diaspora in, 12–13; Caribbean Community (CARICOM) membership, 37, 171; civil unrest in, 106; climate change impact in, 182; credibility and trust building in post-disaster era, 115–16; digital media and post-disaster recovery in, 120–21; earthquakes in, 29, 45, 105–8, 120; election monitoring in, 110; failure of democracy in, 33; hurricane disaster in, 106, 120; information access barriers in, 48–49; innovative practices in, 44–45, 56–57; knowledge accountability and dissemination in, 116–20; knowledge production in, 41; media platforms in, 111–12, 115–20; mediated social capital in post-disaster era, 120–26; political instability in, 29, 34, 56–57; post-disaster media and social capital in, 115–20; post-disaster recovery in, 16, 18–19, 21–22, 29, 31; poverty in, 164; summary of findings in, 201–3; sustainable development in, 4, 185–86; US invasion and occupation of, 108; water management in, 180–81
Haitian Journalists Union Photographers, 116
Haiti Grassroots Watsh (Ayiti Kale Je), 116–20
Haiti Reconstruction Fund (UN), 122–23
Haley, Oretha Castle, 187
Henry, Ariel, 107
hierarchy of needs theory, 194
high-tech expertise, in development discourse, 27–28
historically black colleges and universities (HBCU), knowledge production and capital and, 21–22
Historic New Orleans Collection, 77
Hoover, Herbert, 80
Hurricane Betsy (1965), 80, 89
Hurricane Ida (2021), 16, 31, 70–71
Hurricane Katrina (2005), 16, 29, 31, 70, 80–81, 85, 89
Hurricane Matthew (2016), 106, 120
Hurricane Protection Program, 80
hurricanes, devastation in Louisiana from, 29–32, 80–82
Hutu, genocide committed by, 132–33

identity conflict, genocide and, 144
Indian Removal Act (1830), 74
indigenous African knowledge systems: African diaspora and, 7–9, 12–13; community justice in Rwanda and, 134–35
Indigenous peoples: as chroniclers of environment, 69–70; Ojibwe in Louisiana, 73–75
informal economic structures, 38

information access, knowledge economies and, 45–49
information revolution: development and, 36; education and training and, 39–41; knowledge economies and, 38
information sharing. *See* knowledge sharing
infrastructure: in New Orleans, post-hurricane restoration of, 73–75, 84–90; in post-genocide Rwanda, 133*See also* natural infrastructure movement
in-kind contributions, economic growth and, 38
Innovation Station of the Gulf Coast, 96, 173–74
innovative practices: grassroots coalitions and, 90–97; knowledge economies and, 41–45; sustainable development and, 56–57
The Innovation Station project, 43, 56
institutional economic structures, knowledge economies and, 36–38
inter-African trade, 14
Intergovernmental Panel for Climate Change (IPCC), 9–10, 20, 159, 162
international aid programs: Haitian post-disaster recovery and, 113–14, 118–20; investigation in Haiti of, 121–23
International Criminal Tribunal for Rwanda, 134–35
International Energy Agency (IEA), 12
international law, justice and, 1
International Union for Conservation of Nature, 86
inventions and inventors: Black inventions and inventors, 214–17; overview of, 212–13

Jamaica, climate change policy in, 205
Jefferson, Thomas, 74

Kenya, gender equality in, 174
Keynes, John Maynard, 15–16, 30
Kinyarwanda culture, in post-genocide Rwanda, 145, 151–52, 176
knowledge economies: creation of, 10–11, 195–96; cultural systems and production of, 30; education and training and, 38–41; elements of, 36–49; information access and, 45–49; innovative systems and, 41–45; institutional economic structures in, 36–38; Louisiana and, 72; non-western structures, 124–26; poverty reduction and creation of, 166–69; transitions for sustainability and, 57–59
knowledge production and knowledge capital: accountability and dissemination in Haiti and, 116–20; Black diaspora and, 7, 25–29, 35–36, 195–96; in Caribbean, 44, 172–74; cultural knowledge systems and, 39–41; democratization of, 112–14; development and sustainability and, 3, 15–16, 30–32; disaster recovery and, 2–3, 9–11; gender equality and, 20–21, 170–71, 175–77; in Haiti, 106–10, 120–23; sustainable development and, 82–84
knowledge sharing: Black diaspora and, 96–97; credibility and trust building in post-disaster Haiti, 115–16; digital media and, 111–12; economic growth and, 38; gender equality and, 172–77; history of, 10–11; information access and, 48–49; in Louisiana, 188–89; post-disaster development and, 13–14; in post-genocide Rwanda, 141; in Rwanda, 204
#KotKobPetwoKaribeA, 122–23
Kresge Foundation, 86
Kreyol (Haitian language), social media use of, 122–23

landscape architecture, natural infrastructure and, 87–90
language, gender equality and, 171
La Salle, René-Robert Cavelier, sieur de la, 74

Latin America, economic development in, 30
laughter yoga, in post-genocide Rwanda, 43, 152
Lesser, Eric, 108–9
levees: construction in Louisiana of, 76–80; failure of, 80–82
liberal international economic order (LEIO), limits of, 57–59
Little, Kenneth, 174
local organizations, post-disaster aid programs and, 113–14
Locke, John, 32
Lorde, Audre, 35, 194
Louisiana: Black diaspora in, 12–13; climate change impact in, 65–66, 81–82, 181–82; collective trauma in, 68–69, 83–84, 88–90; community collaboration in, 18–19; emancipation and reconstruction in, 76–78; environmental history of, 73–82; flooding in, 79–82, 87–88, 179–81; forms of disaster in, 69–72; gender equality in, 172–73; hurricane devastation in, 29–32, 80–82; information access in, 46, 48–49; innovative practices in, 41–43, 56; knowledge economies and, 72; natural infrastructure movement in, 87–90; oil industry in, 78; post-disaster recovery in, 3–4, 16, 21, 29, 84–90; poverty in, 164–65; public health crisis in, 177–78; quality of life issues in, 66–68; race-gender wage comparisons in, 165; slavery in, 74–76; summary of findings on, 200–201; sustainable development in, 82–84, 187–89; water resource management in, 177–81
Louisiana Budget Project, 48
Louisiana Environmental Action Network, 92
Lumumba, P. L. O., 14

Maathai, Wangari, 6, 22
Magill, John, 77
manufacturing expertise, in development discourse, 27–28
Marley, Robert Nesta (Bob Marley), 44
Maroons (Indigenous tribes), 76, 187
Marsh, Sarah, 29
Maslow, Abraham, 194
Mason, Sandra, 190
McCalla, Leyla, 186
McCrae, Kiyoko, 186
MediaAlternatif, 116
media platforms: contemporary public sphere and, 110–12; corporate control of, 111; Haitian culture and government and, 109–10; Haitian disaster recovery and, 4, 105–26, 202–3; Haitian post-disaster context for, 114–15; knowledge accountability and dissemination in Haiti and, 116–20; non-western structures for, 124–26; post-disaster aid management and, 113–14; post-disaster Haitian media, 115–20
mental health interventions: in post-genocide Rwanda, 135–38; religious practices and, 142; socio-therapeutic training, 138–41
messaging systems, networking and use of, 112
micro-financing: economic growth and, 38; gender equality and, 176–77
Miller, Portia Simpson, 171
Mirambeau, Gilbert Jr., 122
Mississippi River: drainage and flood control on, 76–78; flooding of, 79–80, 82–83; grassroots coalitions near, 93; New Orleans geography and, 73–75
Mississippi River Gulf Outlet (MRGO), 80
Model of Architecture Serving Society (MASS), 167–68
modernization theories: development and, 34; limits of, 30
Moïse, Jovenel, 107, 110, 120
Moïse, Martine, 107
monetary policy, economic growth and, 35–36
Monsanto, 119
Montalvo, Yeishka, 65–99

Monterey Seafood Watch, 173–74
Mottley, Mia, 171
Muganda (traditional Rwanda practice), 145
multilateral agencies, Haiti's economic reliance on, 110
music, as knowledge capital, 44
Mwaura, Naomi, 174

Nasser, Gamal Abdel, 37
National Seed Service (Haiti), 119
natural infrastructure movement, 84–90
natural resource extraction: climate change, 70; development linked to, 27–28; health risks of, 71–72; poverty reduction and, 168–69*See also* oil industry; petrochemical industries
Ndi Umunyarwanda (I am Rwandan) program, 142, 147–50
Nehru, Jawaharlal, 37
Nelson's West Indian Readers First Primer, 44
neoprene, health risks of, 92
nervous system dysregulation, trauma survivors and, 137–38
Netherlands, water management assistance from, 85–86, 179–80
new international economic order (NEIL), 58–59
New Orleans, Louisiana: collective trauma in, 68–69; drainage and flood control in, 76–78, 85–90; environmental history of, 73–82; flooding in, 79–82, 87–88; geography and infrastructure of, 73–75; Lower 9th Ward flooding in, 80; natural infrastructure movement in, 84–90; post-disaster recovery in, 16; slavery in, 75–76; summary of findings on, 200–201
New York Times, 79, 201
Ngirente, Édouard, 16
Niagara Movement Speech (Du Bois), 13–14
Nicholson, Denise, 46–47
Nigeria, climate change in, 183–84
Nkrumah, Kwame, 37
Nobel Peace Prize, Maathai as recipient off, 6, 22
No Dream Deferred, 188
Non-Aligned Movement, 37, 183–84
non-governmental organizations: aid delivery and anti-poverty initiatives of, 166; Black diaspora reliance on, 29; disaster recovery and, 113; Haitian accountability and knowledge dissemination about, 119–20, 122–23; Haiti's economic reliance on, 110

Obama, Barack, 30, 187
Obotetukudo, Sol, 8–9
oil industry: climate change and, 70; health risks linked to, 71–72; wealth from, 78
Organization of Economic Cooperation and Development (OECD), 184

Paris, Aubrey, 43, 56, 96, 173
parish coalitions, 91–93
participatory action research, community resilience models and, 137–38
Paultre, Andre, 29
Peacebuilding and Reconciliation program (Rwanda), 146
Perkins, Emily, 77
Persadd-Bissessar, Kamla, 171
Petro Caribe, Haitian investigation of aid distribution from, 121–23
petrochemical industries, poverty and proximity to, 71–72
#PetroCaribeChallenge digital network, 106–8, 120–24
#Peyi-lòk (Country shutdown), 123–24
Pichon-Battles, Colette, 94
Pierre, Marc-Henry, 122–23
Pine Island (New Orleans), 73
place-based development, African diaspora and, 27–28
plantation economy, slavery and, 75–76
Pogge, Thomas, 164, 166
political structures: democracy and, 33–34; gender equality and, 171–72, 174–77; genocide and, 132; rebuilding in post-genocide Rwanda, 140–41, 146–49

pollution: innovative systems for reduction in, 42–43; poverty and proximity to, 70–72
post-colonial era: African diaspora knowledge production and, 26; democracy and equitable development in, 32, 35–36; sociopolitical perspectives on, 146–49
"Post-Disaster Futures: Hopeful Pessimism, Imperial Ruination, and La Futura Cuir" (Bonilla), 40–41
post-disaster recovery and development: climate change and, 28–32; control and disbursement of aid and, 113–14; credibility and trust building in Haiti and, 115–16; Haitian media and public sphere and, 114–15; Louisiana case study in, 3–4, 16, 21, 29, 84–90; in Rwanda, 4–5
post-traumatic stress disorder (PTSD): in Louisiana hurricane survivors, 68–69, 83–84, 88–90, 141; in Rwanda, 132–33, 137–38, 141–42
poverty: climate change impact and, 70–72; sustainable development and abolition of, 163–69
Prestage, Jewel L., 22
Prison Fellowship Rwanda (PFR), 146
psychoanalysis: integrative psychotherapeutic methods in Rwanda, 131; socio-therapy as alternative to, 138–39
Psychology Today, 68–69
psychotic disorders, traditional healers and, 142
public health, sustainable development and, 177–81
public sphere: contemporary mediation of, 110–12; Haitian knowledge dissemination in, 108–10; Haitian post-disaster media and, 114–15; knowledge democratization in, 112–14
Putnam, Robert, 109

quality of life, definitions of, 66–68

racial inequality: collective trauma and, 83–84; health inequalities and, 71–72; poverty and, 164–69
radio, in Haiti, 108
Radio Haiti-International, 186
rape crimes, Rwandan community justice and, 134–35
"Rapport de la Commission Sénatoriale Spéciale d'Enquête sur le fonds Petro Caribe Couvrant les Périodes Annuelles Allant de Septembre 2008 A September 2016)" (Report of the Special Senatorial Inquiry Commission on the PetroCaribe funds for the annual periods spanning September 2008 to September 2016), 120–23
Rawls, John, 1
reconciliation, post-genocide recovery and, 142–50
reconciliation villages (Rwanda), 146
REFRAKA (Rezo Fanm Radyo Kominotè Ayisyen), 116
refugee camps, Haiti knowledge accountability and dissemination concerning, 117–20
regional economic development, growth of, 37–38
Reiffers, Jean-Louis, 36
religious practices, Rwandan genocide recovery and, 142
RELX organization, 48–49
resiliency: of African diaspora, 135; community resilience model (CRM) (Rwanda), 135–38; hallmarks of, 84; Louisiana communities' focus on, 72, 82–84; sociocultural practices for, 130–54
resourcing skills, community resilience model, 137–38
restorative justice in Rwanda, 133–35
Ricks, Alan, 168
RISE coalition, 91–93, 200
Roberts, Shearon, 105–26
Rodin, Judith, 68, 82
Rolfes, Anne, 42–43
Rosemblatt, Karin, 30

Rosenstein, Justin, 10
Rwanda: Black diaspora in, 12–13; burial rites in, 152–53; climate change impact in, 182; cultural practices in, 142; development-resiliency balance in, 153–54; economic growth in, 130–31; gender equality in, 149–50, 174–77; genocide in, 131–33; government post-genocide policies in, 142–50; innovative practices in, 43–44; laughter yoga in, 43, 152; mental health interventions in, 135–38; political instability in, 57; post-disaster development in, 4–5; post-genocide recovery in, 16, 19, 21–22, 29, 31; poverty in, 164, 167–68; reconciliation villages in, 146; religiosity in, 142; restorative justice in, 133–35; sociocultural practices for resilience and recovery in, 150–54; sociopolitical perspectives in, 146–49; socio-therapy in, 138–41; summary of findings on, 203–4; sustainable development in, 189; Umuganda program in, 145
Rwanda National Commission, 132–33

"sacrifice zones," 93
Safe Drinking Water Fee (Louisiana), 178
Safe Drinking Water Program (Louisiana), 177–78
SAKS (Sosyete Animasyon Kominikasyon Sosyal), 116
Sankofa beliefs: Akan culture and history of, 160–62; Black diaspora and, 5, 160–62, 184–85; in Louisiana, 187–89; sustainable development and, 20–21, 160–62; water management and, 179–80
Sankofa Community Development Corporation, 187–89
Sankofa Nature Trail and Wetland Park, 188
Santayana, George, 140
September 11, 2001 attacks, 39
Sewerage and Water Board (New Orleans), 77
Singer, Peter, 163–64
singing, post-disaster recovery and, 44
Sir Arthur Lewis Institute of Social and Economic Studies (Jamaica), 205
64 *Parishes* magazine, 187
slavery: impact on Black diaspora of, 20, 32, 39–41, 58–59; in Louisiana, 74–76
social capital: in Haiti, 105, 107–26; knowledge dissemination in public sphere and, 108–10; non-western structures for, 125–26; post-disaster Haitian media and, 115–20, 122–23; sustainable growth in Haiti and, 4
social networks: digital media and, 106–8, 111–12; post-disaster development and, 3–4; in post-disaster Haiti, 120–26*See also* specific networks
sociocultural practices: post-disaster recovery and, 4–5; in post-genocide Rwanda, 150–53
socio-therapeutic training: community resilience model (CRM) (Rwanda), 136–38; mental well-being and, 138–41
somatic interventions: community resilience models and, 137–38; traditional healers and, 142
SOS Journalists Association, 116
South Africa, post-apartheid community justice in, 134, 148–49
Spencer, Herbert, 169
spontaneous movements, community resilience model, 137–38
sports activities, in post-genocide Rwanda, 150–51
State University of Haiti, 116
storytelling, socio-therapy in Rwanda and, 140–41
Sukarno, 37
sustainable development: climate change policies and, 181–84; cultural knowledge contributions to, 17–18, 20–22; development and, 3; economic security and, 197–98; gender equality and, 169–77; goals of, 162–84; in Haiti, 185–86; Haitian

social capital and, 4; information access and, 46–49; knowledge approaches to, 57–59, 82–84; in Louisiana, 82–84, 187–89; Louisiana case study in, 67–68; poverty reduction and, 163–69; public health and, 177–81; in Rwanda, 189; Sankofa model of, 160–62
systems theory: knowledge capital and, 3; sustainable development and, 17–18

tax dividend policy, poverty reduction and, 164, 166
Taylor, Robert, 92
technological innovation: Black diaspora role in, 14; knowledge production and, 30
Theory of Justice (Rawls), 1
Thunbwerg, Greta, 1–2
Tito, Josip Broz, 37
Tocqueville, Alexis de, 74
tracking skills, community resilience model, 137–38
transitional jurisdiction, Rwandan genocide and, 134–35
Treaty of Chaguaramas, 37–38
Trump, Donald, 110
trust-building: democracy and, 34–36; post-disaster Haiti and establishment of, 115–16
Turner, Lauren, 188–89
Tutsi, genocide against, 131–33
Twitter, #petroCaribeChallenge investigation and, 122–23

Ubuntu (radical hospitality), 19
Ugah, Akobundu, 46
Uganda, postcolonial civil conflict in, 147
Umuganda restoration and patriotism program (Rwanda), 145
UN Goals for Sustainable Development, 13, 20, 162–84
United Nations, green energy recommendations, 9–10
United Nations Climate Action Summit, 1–2
United Nations Covenant on Civil and Political Rights, 49
United States: development discourse in, 27–28; invasion and occupation of Haiti by, 108
Unity and Reconciliation of Rwandans, 142–50
Universal Declaration of Human Rights (UDHR), 32
Université d'État d'Haïti (UEH), 116
University of Rwanda: College of Medicine and Health Sciences, 43; innovative practices of, 48
urban blight (New Orleans), hurricane devastation and, 81–82
Urban Water Plan (New Orleans), 88
USAID aid program, Haitian post-disaster recovery and, 119–20
US Department of Housing and Urban Development (HUD), 88
Uwimbabazi, Penine, 145

Venezuela, disaster aid to Haiti from, 120–23

Waggoner, David, 85–86
Waldron-Moore, Pamela, 1–22, 25–59, 159–91, 194–209
Walls Project (Baton Rouge, Louisiana), 42–43, 95–97, 200–201
Warner, Gregory, 174–75
Water Institute of the Gulf of Mexico, 179–80
water management: case studies in, 18–19; public health and, 177–81; sustainable development and, 21
The Water Collaborative of Greater New Orleans (TWC), 85–89, 173, 200–201
wealth distribution: poverty reduction and, 166–69; sustainable development and, 163–69
Webby Award, 205–7
weddings, in post-genocide Rwanda, 151–52
Western development paradigms: educational structures of, 39–41; limitations of, 15–16; social capital in Haiti and, 109–10; traditional environmental knowledge and, 70
WhatsApp, 112

"When the World Was Black" (Bailey), 203
Williams, Helena, 96, 173
Winthrop, Rebecca, 45
women: environmentalists, grassroots coalitions and, 91–97; post-genocide gender equality in Rwanda and, 149–50; sustainable development and role of, 20–21, 56, 169–77
World Bank, 130; Knowledge Economic Assessment, 58–59
World Bank Group, 86
World Economic Forum, 39, 46
World Resource Institute, 86

yellow fever epidemic (New Orleans), 77

Zimbabwe, post-independence community justice in, 134
Zulkifli, Masagos, 159

www.ingramcontent.com/pod-product-compliance
Ingram Content Group UK Ltd.
Pitfield, Milton Keynes, MK11 3LW, UK
UKHW022010240926
8884IPUK00004B/70

9 781807 581428